A PLUME BOOK

THE DYSLEXIC ADVANTAGE

BROCK L. EIDE, M.D., and FERNETTE EIDE, M.D., have a thriving private practice and award-winning blog. The Eides live with their children in the greater Seattle area. Visit dyslexicadvantage.com.

Praise for *The Dyslexic Advantage*

"*The Dyslexic Advantage* is a paradigm-shifting book that captures the remarkable advantages that come with this different style of thinking. This book should be the first people reach for when they want to learn about what it really means to be dyslexic."

—*New York Times* bestselling novelist Vince Flynn

"Any book that focuses on the benefits of having a dyslexic brain should be celebrated. So blow the bugles and toss flowers to herald the arrival of *The Dyslexic Advantage*. This book has some of the best writing on the subject of dyslexia that I have come across. Reading it brought to mind eureka moments."

—Jeanne Betancourt, the author of *My Name Is Brain Brian* and the Pony Pals series

"How did the two of you know me so well? Anywhere I open your book, clarity falls out."

—Henry Winkler, actor and bestselling author of the Hank Zipzer series

"The authors did a great job explaining dyslexia and really hit the nail on the head! As I was reading, it felt like someone was inside of me writing down my innermost thoughts. Gripping, powerful, and insightful—and for dyslexics, immensely validating."

—Nancy Ratey, Ed.M., MCC, SCAC, author of *The Disorganized Mind*

"A fantastic read for anyone who wants to learn about dyslexia. Brock and Fernette Eide have a knack for explaining complex ideas and scientific work in a simple manner *and* offering great insights. This is probably the most helpful material ever published on dyslexia . . . A classic."

—Manuel Casanova, M.D., professor, Gottfried and Gisela Kolb Endowed Chair in Psychiatry at the University of Louisville

"This provocative book explores the misunderstood side of dyslexia that is characterized by skill and talent. The authors focus on what dyslexic people do well. This is a must-read for parents, educators, and people with dyslexia."

—Gordon F. Sherman, Ph.D., executive director of the Newgrange School and Education Center and former president of the International Dyslexia Association

"A compelling call to action." —*Scientific American Mind*

"I recommend it highly!"

—Thomas Armstrong Ph.D., author of *The Power of Neurodiversity*

"Required reading for parents, teachers, dyslexics, anyone with an interest in the enigma of dyslexia."

—Angela Fawcett, emeritus professor at Swansea University

THE DYSLEXIC ADVANTAGE

Unlocking the Hidden Potential of the Dyslexic Brain

BROCK L. EIDE, M.D., M.A.,
AND FERNETTE F. EIDE, M.D.

A PLUME BOOK

PLUME
Published by the Penguin Group
Penguin Group (USA) Inc., 375 Hudson Street, New York, New York 10014, U.S.A. • Penguin Group (Canada), 90 Eglinton Avenue East, Suite 700, Toronto, Ontario, Canada M4P 2Y3 (a division of Pearson Penguin Canada Inc.) • Penguin Books Ltd., 80 Strand, London WC2R 0RL, England • Penguin Ireland, 25 St. Stephen's Green, Dublin 2, Ireland (a division of Penguin Books Ltd.) • Penguin Group (Australia), 250 Camberwell Road, Camberwell, Victoria 3124, Australia (a division of Pearson Australia Group Pty. Ltd.) • Penguin Books India Pvt. Ltd., 11 Community Centre, Panchsheel Park, New Delhi – 110 017, India • Penguin Group (NZ), 67 Apollo Drive, Rosedale, Auckland 0632, New Zealand (a division of Pearson New Zealand Ltd.) • Penguin Books (South Africa) (Pty.) Ltd., 24 Sturdee Avenue, Rosebank, Johannesburg 2196, South Africa

Penguin Books Ltd., Registered Offices: 80 Strand, London WC2R 0RL, England

Published by Plume, a member of Penguin Group (USA) Inc. Previously published in a Hudson Street Press edition.

First Plume Printing, August 2012
10 9 8 7 6 5 4 3 2 1

The Library of Congress has catalogued the Dutton edition as follows:

Eide, Brock, 1962-
The dyslexic advantage : unlocking the hidden potential of the dyslexic brain / Brock L. Eide, Fernette F. Eide.
p. cm.
Includes bibliographical references and index.
ISBN 978-1-59463-079-8 (hc.)
ISBN 978-0-452-29792-0 (pbk.)
1. Dyslexia—Psychological aspects. I. Eide, Fernette, 1961- II. Title.
[DNLM: 1. Dyslexia—psychology. 2. Brain—physiology. 3. Dyslexia—diagnosis.
4. Learning—physiology. WM 475.6]
RC394.W6E33 2011
616.85'53—dc23 2011023471

Printed in the United States of America
Original hardcover design by Eve L. Kirch

PUBLISHER'S NOTE

Neither the publisher nor the author is engaged in rendering professional advice or services to the individual reader. The ideas, procedures, and suggestions contained in this book are not intended as a substitute for consulting with your physician. All matters regarding your health require medical supervision. Neither the author nor the publisher shall be liable or responsible for any loss or damage allegedly arising from any information or suggestion in this book.

While the author has made every effort to provide accurate telephone numbers, Internet addresses, and other contact information at the time of publication, neither the publisher nor the author assumes any responsibility for errors, or for changes that occur after publication. Further, publisher does not have any control over and does not assume any responsibility for author or third-party Web sites or their content.

To Karina (Braveheart): You're an amazing person and an unending source of joy, and we love you with all our hearts.

CONTENTS

ACKNOWLEDGMENTS

We've been truly blessed during the writing of this book by the support and assistance of many remarkable people, whose help we are pleased to acknowledge.

First, we wish to thank our wonderful agent, Carol Mann, who in a very difficult market managed not only to get us a choice of great offers but also helped guide us to the publisher of our dreams. Thanks for watching out for us in so many ways.

And to our team of dreamworthy—but fortunately real—colleagues at Hudson Street Press/Penguin, words are not adequate to express our thanks for your input and support throughout this project. To editor in chief Caroline Sutton, we express our deepest thanks for sharing and supporting our vision, and for assembling such a fantastic team at Hudson Street. To Meghan Stevenson, whose equally astonishing talents as editor and motivational coach of the "tough love" school contributed immeasurably not only to the clarity and structure of the manuscript but also to our inordinate fear of missing deadlines (by too much . . .), you're a true star of your profession, and it has been an honor to work with you. To Courtney Nobile, our lead publicist, and publicity director Liz Keenan, we feel enormously privileged to be backed up by publicists whose enthusiasm is equaled only by their expertise. To Jason Johnson and Eve Kirch, who were responsible for the wonderful design work on the outside and the inside of the book, respectively, thank you for your beautiful and creative contributions. And to John Fagan and Ashley Pattison in marketing, production editor Katie Hurley, copy editor Sheila Moody, and managing editor Susan Schwartz, our deepest thanks for all your work on our behalf. Also, we'd like to thank Bonnie Bader at Grosset & Dunlap at Penguin Young Readers for passing the book on to Henry Winkler so quickly.

We also wish to offer our profoundest thanks to the families and individuals who've come to our clinic and shared their stories and their lives with us. The real work was yours. All we did was listen.

To all those who agreed to be interviewed for this book, to share their stories and expertise, we offer our fullest thanks. The chance to talk to so many accomplished and creative people was one of our most enjoyable professional experiences ever.

Our special thanks to Tom (Thomas G.) West, who's worked harder and longer than anyone to promote the idea that dyslexia has advantages. Ever since we first met Tom he's been unstinting in his kindness and encouragement toward us. Without any question Tom is one of the most generous and selfless individuals we've ever been privileged to know, and the full extent of his impact is only just beginning to be felt.

And to all of our family and friends, old and new, who've become a part of Karina's support community: without your help—emotional, spiritual, professional, and financial—this book could have never been written. Thank you from the bottom of our hearts.

I would now like to turn to . . . *the advantages of the predisposition to dyslexia*. . . . [To] the apparently paradoxical notion that the very same anomalies . . . of the brain that have led to the disability of dyslexia in certain literate societies also determine superiority in the same brains. We can, thus, speak of a "pathology of superiority" without fear of being contradictory.

Norman Geschwind, M.D., "Why Orton Was Right"

PREFACE

In 2004, a top business school in England sent out a press release with the headline: "Entrepreneurs five times more likely to suffer from dyslexia." Its subheading went on to ask, "What makes Sir Richard Branson, Sir Alan Sugar, and Sir Norman Foster special?"[1]

The answer, as revealed in the body of the press release, was that each of these highly successful entrepreneurs also "suffered from dyslexia," a condition researchers at the school found to predispose individuals with dyslexia quite strongly to entrepreneurial success.

Just how successful are the particular entrepreneurs that they cited? At last count, Sir Richard Branson has a net worth of approximately US$4 billion. Alan Sugar, now Baron (Lord) Sugar of Clapton, has a net worth of US$1.2 billion. Norman Foster, now Baron (Lord) Foster of Thames Bank, has a comparatively smaller fortune of "only" US$400 million—though he does have the consolation of being one of the world's most admired and distinguished architects.

In light of the tremendous success enjoyed by these entrepreneurs, it seems rather odd to describe them as "suffer[ing] from dyslexia." Yet as almost anyone with dyslexia can tell you, being dyslexic really can involve a great deal of suffering: like the suffering of constantly failing at skills others master with ease; the ridicule of peers and classmates; or exclusion from classes, schools, or careers one would otherwise pursue. These experiences can all involve sufferings of the cruelest sort. Yet it's equally clear when we examine individuals with dyslexia—when we see how they think and what they can do and the often remarkable persons they become—that in many respects "suffering from dyslexia" is suffering of a most unusual kind.

This book isn't about dyslexia, but about *the kinds of individuals who are diagnosed with dyslexia*. It's about the kinds of minds they have, the ways they process information, and the things they do especially well. It's not a book about something these individuals *have*. It's about who they *are*.

Most books on dyslexia focus on problems with reading and spelling. While these problems are extremely important, they're not the only—or even the most important—things that individuals with dyslexia find critical for their growth, learning, and success.

As experts in neuroscience and learning disabilities, we've worked with hundreds of individuals with dyslexia and their families. In the process we've found that individuals with dyslexia often share a broad range of important cognitive features. Some of these features are learning or processing challenges—like difficulties with reading and spelling, rote math, working memory, or visual and auditory function. But others are important strengths, abilities, and talents; gifts we call the *dyslexic advantage*. While these features differ somewhat from person to person, they also form recognizable patterns—just as the different musical works of Mozart are distinguishable yet recognizably the work of the same composer.

Traditionally, attempts to understand dyslexia have focused almost entirely on problems with reading, spelling, and other academic skills. As a result, little attention has been paid to the things individuals with dyslexia do especially well—particularly once they've become adults. In our opinion, this is a grave mistake. Trying to understand what dyslexia is all about while overlooking the talents that mature individuals with dyslexia characteristically display is like trying to understand what it's like to be a caterpillar while ignoring the fact that caterpillars grow up to be butterflies.

As we'll show you in this book, the brains of individuals with dyslexia aren't defective; they're simply different. These wiring differences often lead to special strengths in processing certain kinds of information, and these strengths typically more than make up for the better-known dyslexic challenges. As we'll show you in this book, by learning how to recognize, nurture, and properly use these strengths, individuals with dyslexia can be helped in their efforts to achieve success and personal fulfillment.

There are two big differences between the traditional view of "dyslexia" and the one we'll present in this book. First, we don't see the reading, spelling, or other academic challenges associated with dyslexia as the result of a "disorder" or a "disease." Instead, we see these challenges as arising from a different pattern of brain organization—one whose chief aim is to predispose dyslexic individuals to the development of valuable skills. When dyslexia is viewed from this perspective, we can see that the strengths and challenges that accompany it are like two sides of the same neurological coin. In this book, we'll identify these advantages, describe how they can be used, and explain why we believe that they—rather than challenges with reading and spelling—should be seen as dyslexia's true "core features."

Second, unlike most books on dyslexia this book won't focus solely on making individuals with dyslexia into better readers. Instead, it will focus on helping them become better *at "being dyslexic."* While reading instruction changes certain brain features, it doesn't change all the things that make dyslexic brains different from nondyslexic ones. However, this is a good thing, because dyslexic brains aren't *supposed* to be like everyone else's. Dyslexic brains have their own kinds of strengths and benefits, and these advantages should be recognized and enjoyed. Our goal is to help individuals with dyslexia recognize these many wonderful advantages, so they can enjoy the full range of benefits that can come from having a dyslexic brain. The first step in achieving this goal is to help them think more broadly about what it really means to "be dyslexic," by expanding the concept of "dyslexia" so that it no longer means only challenges but also includes important talents.

The best way to broaden our view in this fashion is to look not just at the things that individuals with dyslexia find challenging but also at the kinds of things that they often do especially well. One obvious way to do this is by studying people who've excelled at "being dyslexic." Most instructional books or DVDs on topics like playing sports or musical instruments, cooking, or speaking foreign languages have one thing in common: they feature expert practitioners sharing and modeling tips and strategies they've personally found useful. Since this is a book about how to excel at "being dyslexic," we'll share lots of stories, tricks, and pointers from dyslexic individuals who've enjoyed success in their own lives. While not every individual with dyslexia will succeed in precisely the same ways as these talented individuals, anyone with a dyslexic processing style can benefit from their insights and from studying the strategies they've used.

In the early chapters of this book, we'll describe how dyslexic brains differ from nondyslexic ones. Then we'll devote five chapters to each of the four dyslexia-associated strength patterns we've found to be common in individuals with dyslexia. We've called these patterns the *MIND strengths* to make them easy to remember: Material reasoning, Interconnected reasoning, Narrative reasoning, and Dynamic reasoning. These strength patterns are not meant to be rigid or watertight categories but to be helpful ways of thinking about and understanding dyslexic talents. While none of the MIND strengths is exclusive to individuals with dyslexia, each is linked to particular cognitive and structural brain features common in individuals with dyslexia. As you read these chapters, please remember that while individuals with dyslexia share many features in common, each is also unique. Dyslexic processing isn't caused by a single gene, so different individuals with dyslexia will show different patterns of strengths and challenges. Very few will show all the MIND strengths, but essentially all will show some. After discussing the MIND strengths, we'll conclude with several chapters of practical advice describing how individuals with dyslexia can profit from their dyslexic advantages both in school and at work.

We hope that this book will provide a resource and an encouragement for those who haven't yet fully learned the many wonderful advantages that can come from "being dyslexic."

PART I

A Matter of Perspective

CHAPTER 1

A New View of Dyslexia

Throughout his school career, Doug struggled with reading and writing. He flunked out of community college twice before he finally gained the skills he needed to earn his college degree. Today he's the president of a highly successful software firm that he founded a decade ago.

When Lindsey was young all her teachers called her slow. Although she worked desperately to learn to read and write, she was one of the last in her school class to master these skills. Recently Lindsey graduated from college, earning the top prize in her school's highly competitive honors program. She's now enrolled in a prestigious graduate program studying psychology.

Pete's elementary school teachers told his parents he was borderline mentally retarded and emotionally disturbed. They also told them that they couldn't teach him to read or write. However, with intensive one-on-one instruction, Pete learned to read and write well enough not only to attend college but also to go on to law school. Pete eventually used his legal training to represent another individual with dyslexia before the Supreme Court, winning her case 9–0 and radically redefining the rights of students with special educational needs.

Doug, Lindsey, and Pete are all dyslexic, and they're also all exceptionally good at what they do. As we'll show you in this book, these facts are neither contradictory nor coincidental. Instead, Doug, Lindsey, and Pete—and millions of individuals with dyslexia just like them—are good at what they do, not *in spite of* their dyslexic processing differences, but *because* of them.

This claim usually provokes surprise and a flurry of questions: "Good because of their dyslexia? Isn't dyslexia a learning disorder? How could a learning disorder make people good at anything?"

The answer is, a learning disorder couldn't—if it were *only* a learning disorder. But that's just our point, and it's the key message of this book. Dyslexia, or the *dyslexic processing style*, isn't just a barrier to learning how to read and spell; it's also a reflection of an entirely different pattern of brain organization and information processing—one that predisposes a person to important abilities along with the well-known challenges. This dual nature is what's so amazing—and confusing—about dyslexia. It's also why individuals with dyslexia can look so different depending upon the perspective from which we view them.

Look first at individuals with dyslexia when they're reading or spelling or performing certain other language or learning tasks. From this perspective they appear to have a learning disorder; and with respect to these tasks, they clearly do. Now look at these same individuals when they're doing *almost anything else*—particularly the kinds of tasks they excel at and enjoy. From this new perspective they not only cease to look disabled but they often appear remarkably skilled or even specially advantaged.

This apparent advantage isn't just a trick of perception—as if their strengths seemed large only in contrast with their weaknesses. There's actually a growing body of evidence supporting the existence of a *dyslexic advantage*. As we'll discuss throughout this book, many studies have shown that the percentage of dyslexic professionals in fields such as engineering, art, and entrepreneurship is over twice the percentage of dyslexic individuals in the general population. Individuals with dyslexia are also among the most eminent and creative persons in a wide variety of fields, like entrepreneur Richard Branson, singer-songwriter John Lennon, paleontologist Jack Horner, financial services pioneer Charles Schwab, inventor Dean Kamen, architect

Richard Rogers, attorney David Boies, novelist Vince Flynn, computer pioneer Bill Hewlett, actor Anthony Hopkins, painter Chuck Close, cell phone pioneer Craig McCaw, and filmmaker Bryan Singer.

Importantly, the link between dyslexic processing and special abilities isn't visible only among superachievers. You can prove this for yourself by performing a simple experiment. Next time you run across an unusually good designer, landscaper, mechanic, electrician, carpenter, plumber, radiologist, surgeon, orthodontist, small business owner, computer software or graphics designer, computer networker, photographer, artist, boat captain, airplane pilot, or skilled member of any of the dozens of "dyslexia-rich" fields we'll discuss in this book, ask if that person or anyone in his or her immediate family is dyslexic or had trouble learning to read, write, or spell. We'll bet you dollars for dimes that person will say yes—the connection is just that strong. In fact, many of the most important and perceptive experts in the field of dyslexia have remarked on the link they've seen between dyslexia and talent.

Now, would these connections be possible if dyslexia were *only* a learning disorder? The answer, clearly, is no. So there must be two sides to dyslexia. While dyslexic processing clearly creates challenges with certain academic skills, these challenges are only one piece of a much larger picture. As we'll describe throughout this book, dyslexic processing also predisposes individuals to important abilities in many mental functions, including:

- three-dimensional spatial reasoning and mechanical ability
- the ability to perceive relationships like analogies, metaphors, paradoxes, similarities, differences, implications, gaps, and imbalances
- the ability to remember important personal experiences and to understand abstract information in terms of specific examples
- the ability to perceive and take advantage of subtle patterns in complex and constantly shifting systems or data sets

While the precise nature and extent of these abilities varies from person to person, there are enough similarities between these strengths to form a recognizably related set, which can legitimately be referred to as *dyslexia-*

related abilities or a *dyslexic advantage.* Ultimately, that's what this book is about: the remarkable abilities that individuals with dyslexia commonly possess—abilities that appear to arise from the same variations in brain structure, function, and development that give rise to dyslexic challenges with literacy, language, and learning.

In this book we'll argue for a radical revision of the concept of dyslexia: a "Copernican revolution" that places abilities rather than disabilities at the center of our ideas about what it means to be an individual with dyslexia. This shift in perspective should change not only our thinking about dyslexia but also the ways we educate, employ, and teach individuals with dyslexia to think and feel about themselves, their abilities, and their futures.

Please understand that we're not trying to downplay the hardships that individuals with dyslexia can experience or to minimize their need for early and intensive learning interventions. We are simply trying to expand the view of dyslexic processing so that it encompasses both the challenges that individuals with dyslexia face and the abilities they commonly demonstrate. This broadened perspective can be illustrated using the following analogy.

A Tool Discovered

Imagine you live on a remote island and you've never had contact with the people or products of the outside world. One morning as you walk along the beach you spy a shiny cylindrical tube half buried in the sand. You pick it up, clean it, and carefully examine it. With growing excitement you realize it's a product of human design, but what it is, or what it's for, you can't immediately decide.

As you inspect the tube you find that it's roughly as long as your arm and as heavy as a fist-sized stone. It's also gently tapered so that one circular end is nearly twice as wide as the other. As you turn the tube to inspect its large end, you notice that the light shines from it with special intensity. When you bring this end toward your eye, you discover that the light isn't just bouncing off the tube's end—it's shining either from it or through it. You peer cautiously through this large end, and after a moment's adjustment you begin to see a familiar yet marvelously transformed image: it's a lovely, delicate miniature of the beach stretched out in front of you. With awe and astonishment you realize what you've discovered: a remarkable device for making things look small!

Well, yes and no. . . .

Like a telescope, the concept of dyslexia is a human invention; and like a telescope it can either expand and clarify our view of individuals who struggle to read and spell or, used "the wrong way around," it can cause our view of these individuals to shrink. Unfortunately, this "diminishing effect" is just what we believe has happened with the way the concept of dyslexia has been used.

How the "Narrow View" Became the Primary View

Surprisingly, given how common we now know dyslexia to be,[1] the first clear description of an individual with dyslexia appeared in the medical literature just over a century ago. In 1896, British ophthalmologist W. Pringle Morgan described a fourteen-year-old boy named Percy, who despite receiving seven years of "the greatest efforts . . . to teach him to read," could read and spell only at the most basic level, even though his schoolmasters believed he was "the smartest lad in the school."[2]

It was through this case that the concept of dyslexia was first developed: the idea that there exists a distinct group of individuals who—though clearly intelligent—learn and process certain kinds of information very differently from their nondyslexic peers. Historically, the processing features most commonly associated with dyslexia are difficulties with reading and spelling, though, as we'll see later, other challenges with language and learning are also common in individuals with dyslexia.

While the development of this concept has been tremendously useful, we believe its true worth has never been fully realized because, like the telescope in our example, one critical question has been overlooked: *How should this "telescope" be used?* Should it be used as a tool to narrow our view solely to literacy, language, and learning difficulties? Or should it be "turned around" so we can see *all* the learning and processing features of this amazing group of individuals: not just in literacy and language but across the whole range of their activities—strengths as well as challenges—and throughout their entire life span?

Since this question has largely gone unasked, challenges with literacy, language, and other aspects of learning have remained the almost exclusive focus of dyslexia research and education. As a result, "dyslexia" has come to be seen as essentially synonymous with those challenges. This perspective is reflected in current definitions of dyslexia. In the United States, the most widely used definition was developed by the National Institute of Child Health and Development (NICHD) and subsequently adopted by the International Dyslexia Association (IDA). It reads:

> Dyslexia is a specific learning disability that is neurological in origin. It is characterized by difficulties with accurate and/or fluent word recognition and by poor spelling and decoding abilities. These difficulties typically result from a deficit in the phonological component of language that is often unexpected in relation to other cognitive abilities and the provision of effective classroom instruction. Secondary consequences may include problems in reading comprehension and reduced reading experience that can impede growth of vocabulary and background knowledge.

In the terms of our telescope analogy, this is clearly a "shrinking" perspective since it narrows our view of dyslexic processing to the challenges experienced by individuals with dyslexia, and it does nothing to expand our view of their skills or capacities. From this perspective, dyslexia is merely:

- a *learning disability*
- characterized by *difficulties*

- resulting from *deficits*
- that produce *secondary consequences*
- and additional *impediments*

No wonder people have such a negative view of dyslexia!

But is there really any reason to believe that this definition tells us everything we need to know about individuals with dyslexia? In a word: no. Habit alone has led us to assume that the first use we discovered for this "telescope" is the only one, and habit alone keeps us from discovering other—and better—uses.

Ultimately, we've recognized the phenomenon of dyslexia but missed its significance—like archaeologists who've discovered a vast and elaborately carved gate but become so engrossed in its study that we've failed to realize that a magnificent city must lie buried nearby. Because we first recognized dyslexia as a *learning disorder* rather than a *learning* or *processing style*, we've paid little attention to whether dyslexic processing might also create talents and abilities. However, as we'll show you in the next chapter, the talents and benefits that are associated with dyslexic processing can be easily observed once we recognize that dyslexia can be viewed from two different perspectives.

CHAPTER 2

Dyslexia from Two Perspectives

To demonstrate the enormous difference that results when dyslexia is viewed from these two different perspectives, we'd like to introduce you to a family we've been privileged to meet through our work.

We first met Kristen when we spoke to a parents' group about the challenges we often see in very bright children. After our presentation, Kristen introduced herself and told us about her son. Christopher was in third grade, and he'd recently shown very broad gaps in his performance on the different subtests used in IQ testing. While he'd scored well on tests measuring verbal and spatial reasoning, his performance was weaker on tests measuring processing speed and working memory (or "mental desk space," which we'll discuss later). Kristen wanted to know if we could tell her anything about children who show such a pattern.

We told Kristen that we often see this pattern in bright young boys, whom we've affectionately dubbed our "young engineers." Although many of these boys show strong interest in verbal subjects like history, mythology, fantasy literature, role-playing games (including game creation), reading or being read to, and even storytelling or creation of imaginary worlds, they typically show their keenest interest in spatial or mechanical activities like building, designing, art, inventing, electronics, computing, and science. We told her that many of these boys struggle with handwriting, written expression, spelling, and initially with reading (especially with oral reading fluency). Often these boys are persistently slow readers, and a smaller number show challenges with oral language expression, like word retrieval, or difficulty putting their thoughts into words. Many also have a clear family history of dyslexia or many relatives who've excelled as adults in occupations requiring spatial, mechanical, or higher mathematical skills.

Kristen at first appeared surprised by our remarks—almost taken aback—and we wondered if we'd missed our mark. But as we finished, she slowly smiled and said, "Let me tell you about my family. . . ."

One Dyslexic Family: The Narrow Perspective

Kristen's son, Christopher, showed his first dyslexia-related challenges quite early in development. Like many dyslexic children, he was slow to begin speaking (his first words came shortly after his second birthday) and slow to combine words into sentences. As a preschooler his speech was often unclear, and he struggled to find words to express his thoughts. He often subtly mispronounced words and confused similar-sounding words like *polish* and *punish.* Despite the fact that he could identify the numbers zero through ten before his second birthday, Christopher couldn't learn his letters until he was almost five years old. In school, he was much slower learning to read and write than most of his classmates, and he also had great difficulty memorizing math facts, despite a strong grasp of number concepts and math reasoning.

Christopher received special testing and was referred to several learning specialists who helped him with reading, handwriting, speech articulation, and word retrieval. He's currently in fourth grade, and though his reading accuracy has improved, he still struggles with slow written work production, messy handwriting, and spelling.

Kristen, too, showed many signs of dyslexia early in her life. She was very slow in learning to read, and according to her parents she still struggled with basic phonetic decoding as late as fourth grade. Like Christopher she also made frequent word substitutions (like *peaches* for *pears*), struggled to get her thoughts down in writing, and spelled very poorly. Kristen also had a weak memory for auditory or verbal sequences—like phone numbers or word spellings—and she struggled to master abstract verbal concepts that she couldn't easily picture.

Kristen recalled that during her early years in school she was often "incredibly bored" and "couldn't stand desk work." She found listening to lectures on abstract subjects especially difficult. For much of middle school and the early years of high school her grades were poor, and she came close to failing out. Then it finally hit her that time was flying by and that "if she wanted to get anywhere in life" she'd have to go to college; so she buckled down and was able to raise her grades sufficiently to get into a state university.

Kristen initially planned to major in sociology or psychology—the subjects she found most interesting; but she quickly realized, "If I had to read or write for my degree, I wouldn't make it through." So instead she majored in interior design, and after earning her degree she went to work for a large design firm.

Kristen's father, James, doesn't remember any unusual difficulties learning to read as an elementary student in the 1930s. However, throughout his life he's shown a persistent discrepancy between his high intellectual ability and his difficulty learning from text, which is characteristic of individuals with partially compensated dyslexia. He's always been a slow reader, has never read for pleasure, and according to his family was able to succeed in high school and college largely because his childhood sweetheart—and now wife of nearly sixty years—Barbara, helped him do the reading for his coursework. She still helps him with business-related reading.

Handwriting, spelling, and written expression also troubled James throughout his education, and they've remained tough for him to this day. Kristen fondly recalls the time when she made an imaginary diner from a cardboard box and asked her father to help her spell *restaurant* above its door. He thought for a moment, then said, "You should call it a *café* instead—that's a much nicer word." James also had difficulty remembering math facts like the times tables (a problem he's still never entirely overcome);

remembering math equations and certain rules and procedures; taking notes during lectures; grasping verbal (and especially abstract verbal) concepts; mastering a foreign language; switching attention from one subject to another as required at school (despite good prolonged attention for his preferred activities); and feigning interest in almost everything they were trying to teach him in school, with the exception of his advanced science courses.

This is the *narrow* view of Christopher, Kristen, and James that emerges when we focus exclusively on their dyslexia-related challenges. Now let's see how they appear when we broaden our focus to look instead at their strengths.

One Dyslexic Family: A Second Look

Christopher, at age nine, already shows many of the talents and abilities that are common among individuals with dyslexia. First, he shows very strong three-dimensional spatial abilities, which have revealed themselves in several ways since early in life. For example, when Christopher was only three, his family was staying at a large hotel, which had been formed by combining several older and very different structures into one enormous complex. After checking in at the front desk, the family walked for several minutes through a confusing labyrinth of passages to reach their room. Once there, they deposited their bags and walked out again in search of dinner. When they returned to the hotel several hours later, Christopher announced that he would lead the family back to their room. To his parents' astonishment, he did so without a single hesitation or mistake.

Christopher's spatial abilities have also revealed themselves in his persistent love of building. Although he enjoys using just about any kind of building material, LEGOs are a special favorite, and he often spends hours using them to build complex and unique designs in a room in his house devoted solely to this purpose. He also displays a passionate interest in science and how things work.

Christopher also shows many impressive verbal strengths, despite some persistent focal language challenges. He's always had a great love of stories, and even before he could speak he would listen with rapt attention to the reading of lengthy stories, like *The Velveteen Rabbit*. Now that his reading skills have improved he's become a voracious reader, and he reads for both entertainment and information. Despite his difficulties with verbal output, Christopher's verbal scores were some of his strongest on his IQ testing.

Kristen also appears much different when we look at her from this "broadened" perspective. Although her school career was marked by difficulty remembering many kinds of abstract verbal facts, Kristen's memory is remarkably good in other ways. By mentally "tapping out" numbers on an imaginary keyboard, Kristen can recall a long list of phone numbers, including those of many places where she's worked or lived and friends' numbers stretching back into childhood.

In addition, Kristen, like Christopher, has a very strong spatial sense and can quickly and indelibly learn her way around new environments. She also has a phenomenal visual memory for people and places from her past and can still easily "see" where she sat in all her school classes as a child, who sat around her, what many of them wore, and how the walls of her classrooms were decorated.

Many—perhaps most—of Kristen's memories have a strong contextual, personal, or "episodic" element, involving elements of past experience. Kristen experiences these memories like dramatic scenes playing out in her mind. They portray information about where she first encountered each fact, object, individual, fashion, song, or other remembered item, including whom she was with, what she saw or heard, and how she felt. Kristen also experiences a similar sensory-immersive experience of sounds, colors, touch sensations, and emotions when she reads or hears stories. As we'll see in later chapters, this type of vivid *episodic* memory is extremely common in individuals with dyslexia, and it is often accompanied by weaknesses in abstract verbal, or *semantic*, memory.

As with many of the strong "personal" learners with whom we work, Kristen also finds learning to be a very personal, almost intimate, act. During her years in school, this made her learning highly dependent upon her relationship with her teachers and her interest in her course materials.

While Kristen's memory skills are impressive, they did little to help her with her schoolwork—though, channeled properly, they clearly could have. Instead, her vivid recall of personal experiences often created a powerful inducement to daydream. As a result, it took a great deal of interesting "outside" stimulation from her teachers to hold her attention.

Eventually, however, these cognitive traits became the foundation for Kristin's highly successful career. After finishing college, Kristen went to work for a firm that designed and furnished office spaces. She quickly became one of the most productive design and sales representatives in this nationwide operation. Kristen credits much of her professional success to her spatial and personal memory skills, which allow her to imagine how interior spaces will look when changed in various ways. She also finds that her restless energy, drive, and dislike of desk work—all of which made it so hard for her to sit passively in class every day—are ideally suited for a job that requires visits to construction sites, suppliers' showrooms, clients' offices, and frequent phone calls to troubleshoot issues.

Kristen's father, James, also discovered that many of the cognitive features that troubled him in school became keys to his success in the working world. While inside the classroom he showed few early signs of promise, outside of school he displayed a remarkably bright and precocious mind. At age six he built his first radio-controlled boat, which he designed to include a special compartment to carry his lunch. He spent much of his time "taking things apart" to see how they worked, and his interest in electronics was further piqued when an electrician visited his home to install a new stove and took the time to demonstrate his tools and techniques.

James also developed an interest in magnetism. When he was a student during World War II and dozens of sand buckets were brought to his school for fire safety, James demonstrated that the sand was full of iron by running a magnet through the buckets. Rather than receiving encouragement for his interest in experimental science, James was disciplined for "playing" in the sand.

In tenth grade James finally found a science teacher who could answer his insightful questions and who was able to lead him into new and deeper areas of interest. Chemistry and physics became special passions, and he reveled in the pleasure of having a teacher who saw him as especially promising. In response, James not only completed all the required reading, but he also labored through several more advanced books. Outside of school, James strengthened his knowledge of electronics by working for the electrician who'd earlier befriended him. During the summer after his junior year of high school, James put this knowledge to work by building a commercial-grade AM radio station, which he sold to a local entrepreneur.

After earning his degree in physics at Reed College, James went to work for Battelle Memorial Research Institute in Richland, Washington. There he quickly distinguished himself as a talented and creative inventor. He received his first patent for an electron beam welder, then he followed up with a steady stream of inventions that he created to solve problems for clients.

However, James's most famous invention had its origin not in a client's problem but in one of his own. James has always loved classical music, and he loves to play his favorite recordings again and again and again. Back when all his music was stored on LP records, he was driven nearly crazy by the hisses, scratches, and skips that accumulated when he played his favorite records repeatedly. Seeking to eliminate the wear that came from repeated physical contact between a stylus and a grooved vinyl record, James imagined a system where an optical reader would detect digital information embedded on a small plate, with which it never made physical contact. Over the next several years, James invented the seven components that together became known as the compact disc system. The impact of this invention on data storage and retrieval—not just for music but for all types of information—has been profound. In fact, you'll find James T. Russell's compact disc system on many lists of the most important inventions of the twentieth century.

Now nearing eighty, James is still active as an inventor. He has nearly sixty U.S. patents to his name and four more currently in process. He continues to work nine hours each day in the laboratory he's built in his home, and he's confident that his best inventions are still to come.

Dyslexia and Talent: An Essential Relationship

The lives of James, Kristen, and Christopher—though in some ways unique—display many of the features we commonly observe in individuals with dyslexic processing styles. In fact, it was seeing these patterns repeated in the dyslexic families with whom we work that convinced us that certain strengths are as much a part of the dyslexic profile as challenges in reading and spelling.

Please notice that we're *not* saying merely that individuals with dyslexia can be talented *in spite of* their dyslexia—like Lance Armstrong overcoming cancer to become a seven-time Tour de France winner or Franklin D. Roosevelt overcoming polio to become president of the United States. Instead, we're claiming that certain talents are as much a part of dyslexic processing as the better-known challenges—that the strengths and the challenges are simply two sides of the same neurological coin.

We can explain what this connection is like using an example from the sport of baseball. Consider the following players:

Barry Bonds
Willie Mays
Frank Robinson
Harmon Killebrew
Jim Thome
Mickey Mantle
Willie McCovey
Hank Aaron
Ken Griffey, Jr.
Mark McGwire
Rafael Palmiero
Mike Schmidt
Jimmie Foxx
Babe Ruth
Sammy Sosa
Alex Rodriguez
Reggie Jackson
Manny Ramirez
Frank Thomas

If you're at all familiar with baseball, you'll recognize at least a few of these names, and if you're a real fan, you'll recognize them all. These are some of the greatest stars who've ever played the game. Consequently, you may be surprised to learn that another thing they have in common is that all are among the top one hundred all-time "leaders" in striking out while at bat!

This seems like a rather unwelcome distinction, since striking out is unquestionably a kind of mistake—like misspelling a word or misreading a sentence or writing illegibly. If you knew nothing more about these players than that they'd struck out more than almost all the other batters in major league history, you'd probably conclude that they were poor players. You might even conclude that they shared a kind of hitting disability, like "dysfunctional batting syndrome" or "contact deficit disorder."

However, one thing is certain: if you didn't also know that these players are the top nineteen home-run hitters of all time, you'd have a highly incomplete—and seriously misguided—impression of their value and ability as players. Although striking out is clearly undesirable in itself, when viewed in the context of the game as a whole, we discover that even the greatest players strike out a lot if they swing hard enough and often enough to hit a lot of home runs. Since avoiding strikeouts isn't as important in baseball as scoring runs—which these big-swinging home-run hitters did extremely well—the list of strikeout leaders turns out, rather surprisingly, to be a list of some of baseball's greatest winners, not its losers.

This relationship between home runs and strikeouts is a lot like the connection between the strengths and challenges in dyslexia. The "home runs" that dyslexic brains have been structured to "hit" are *not* perfect reading and spelling but skills in other kinds of complex processing that we'll discuss throughout this book. And it's because dyslexic brains have been organized to make these "home runs" possible that they're also at higher risk for "striking out" when they try to decode or spell words. The weaknesses are simply the flip side of the strengths.

Discovering these strengths—and how they can be used to help individuals with dyslexia find success in the classroom and the workplace—is what this book is all about. We'll begin our search for these strengths in part 2, where we'll look at how dyslexic brains have been shown to differ from nondyslexic ones, both in how they work and in how they're structured. As you'll see, these differences aren't responsible for just the challenges associated with dyslexia; they're the source of important advantages as well.

PART II

How Dyslexic Brains Differ

CHAPTER 3

Differences in Information Processing

Before we discuss some of the key differences between dyslexic and nondyslexic brains, let's take a moment to look at the behaviors or "symptoms" associated with dyslexia that these differences are intended to explain. We'll begin by listing dyslexia-associated challenges, because that's what experts have usually focused on.

The first challenges that many children with dyslexia display involve language. Some dyslexic children are late-talking. Many others alter, leave out, or reverse word parts (e.g., *berlapse/relax, wold/world, pasghetti/spaghetti*) or even invent their own unique words for things. Dyslexic individuals often struggle to retrieve words from memory, and they may be slow to master the use of tenses, cases, pronouns, or other grammatical rules.

In the preschool or early elementary years dyslexic children often struggle to perceive rhymes, and many have difficulty learning to break words into their component sounds (e.g., *c-a-t*) or learning the names and sounds of different letters. Early in school most—but not all—dyslexic children will show obvious struggles with reading and spelling. (A few, whom we've elsewhere called *stealth dyslexics*, have problems so subtle or "stealthy" that they evade early detection and often only come to attention later for problems with writing or underperformance.)[1]

Many dyslexic students also show problems with handwriting and written expression; basic arithmetic and rote memory for math facts; processing speed; motor coordination; mishearing and difficulty hearing in background noise; visual function for near work; following directions; keeping information in their mind (working memory); mastering procedures; planning and organization; error detection; time awareness and pacing; sequencing; and mental focus and attention. Individuals with dyslexia may also show subtle difficulty in learning the rules that describe how words work together in groups (grammar and syntax). These latter problems are often recognized only in the middle to upper elementary grades when students are asked to express more complex ideas and to read or write more structurally complex sentences.

This list of challenges may make it seem that individuals with dyslexia face a whole host of different "problems." Actually, all these findings can be traced back to a small number of variations in brain structure and function. It's only because these variations occur in basic processing systems that are used for many different functions that they can give rise to such a wide variety of "symptoms."[2] For most individuals with dyslexia, it's likely that only a very few underlying variations are responsible for all their dyslexia-associated findings. This is true even for many of the individuals with dyslexia who receive multiple diagnoses, like dyslexia "plus" attention deficit disorder, dyspraxia, developmental coordination disorder, or auditory processing disorder.

In this chapter and the next, we'll look at four important brain variations that have been found to be associated with dyslexia. We'll examine how these variations in brain function or structure may be responsible for both the dyslexia-associated challenges we've listed and the dyslexic advantages we'll discuss in later chapters. Let's begin, in this chapter, by looking at two dyslexia-associated variations in information processing (or *cognition*).

Phonological Processing

The first pattern we'll discuss is a variation in the brain's *phonological* (or "word sound") *processing system*. This system is used to process phonemes, the basic sound components in words. English has approximately forty-four phonemes, and just as the letters of our alphabet can be strung together to form printed words, phonemes can be strung together to form all the spoken words in English.

For the past thirty years most reading specialists have favored the *phonological impairment theory* as the most likely explanation of the brain basis of dyslexic reading and spelling problems. Think back for example to the definition of dyslexia cited in chapter 1, which states that dyslexia-associated reading and spelling difficulties "typically result from a deficit in the phonological component of language."

There are several good reasons for believing that phonological impairments play a key role in causing dyslexic reading and spelling challenges. Problems with phonological processing have been found in at least 80 to 90 percent of individuals with dyslexia, and they can clearly contribute to many of the challenges mentioned earlier in this chapter. The role that phonological processing impairments play in the reading and spelling challenges many individuals with dyslexia display has been especially well worked out. While we'll save a fuller discussion of this role for our chapter on reading, there are several key points we should mention here.

The phonological processing system plays a key role in analyzing and manipulating the sound structures of words. Many of these functions are important for matching word sounds and the letters used to represent them—that is, for mastering the rules of phonics which underlie decoding (or sounding out words) and encoding (or spelling words). Two of the most important phonological processing (or *phonological awareness*) tasks underlying these skills are *sound segmentation* (or the ability to split incoming words into their component sounds) and *sound discrimination* (or the ability to distinguish word sounds from one another). Most individuals with dyslexia struggle with one or both of these tasks and as a result have difficulty mastering the basic skills underlying reading and spelling.

Even though phonological processing involves low-level or fine-detail language processing—that is, processing of the most basic building blocks of language—it forms the foundation for the entire language structure and supports many of the higher language functions. That's why severe problems with phonological processing can cause difficulty at all levels of language, such as mastering word meanings, learning how words interact when used in groups (that is, grammar and syntax), and understanding how words work together to form "discourse-level" messages like paragraphs or essays. When the higher-order language problems resulting from phonological impairments are severe, they are called "specific language impairment," but the underlying process remains the same.

The phonological processing system also plays an important role in many attention-related functions, including working memory and executive functioning. *Working memory* is the kind of short- to intermediate-term memory that helps us "keeps things in mind" for active conscious processing—very much like the random-access memory or RAM on your computer. The phonological processing system forms a *phonological loop* (or short-term memory tracing) that keeps auditory-verbal information alive in active working memory until it can be processed, organized, and put to use.

When auditory-verbal working memory is limited (or has too short a functional span), the brain may fail to finish all the processing it needs to perform before this "internal speech tracing" fades away. The result is *working memory overload*, which causes symptoms like inaccurate language processing, slower language-based learning, problems with organization and task management, and the appearance of inattention during difficult work. Working memory overload resembles what happens when you try to run a memory-intensive software program on a computer with too little keyboard memory. At first the program runs more slowly; then it begins to flash error messages; and finally it jams up completely.

Problems with working memory overload are very common in dyslexic students. They often first appear during the early elementary years, when complex tasks like reading, writing, and math are first introduced; peak again during the middle elementary years, when organization and study skills are first stressed; then cause another peak of challenges during middle and high school when language and organizational demands become even more complex.

Importantly, working memory also plays a key role in other aspects of attention or "executive function" like organization, planning, implementation, and oversight of tasks. That's why when working memory is limited due to problems with phonological processing, students can experience a whole range of challenges with attention. Often such students are diagnosed with inattentive ADHD.

Problems with phonological processing are usually attributed to structural variations in the brain's left hemisphere, particularly in the language areas of the left temporal lobe. The precise nature of these variations isn't fully known. Some researchers believe they are caused by alterations in processes that take place very early in development, when brain cells organize themselves into functional networks. Because the networks don't form in a well-integrated fashion, the processing of phonological information is impaired.[3] Other researchers have proposed that these impairments are caused by difficulties in learning rule-based procedures or by inherited variations in the structure of the brain's circuitry. We'll discuss these hypotheses in more detail in the rest of this chapter and the next.

For the moment, however, let's focus on the key question of whether phonological processing impairments by themselves seem capable of causing all the challenges and strengths associated with dyslexia. It shouldn't take us long to see that they cannot. For example, there's no direct relationship between poor phonological processing and common dyslexia-associated difficulties like problems with finger coordination for handwriting, eye movement control for reading, or speech muscle control for speech articulation. Even more importantly, phonological processing impairments provide no explanation for the kinds of dyslexic advantages or dyslexia-associated processing strengths that we saw in Kristen, Christopher, and James—such as their strong mechanical and spatial abilities or their strengths in spotting unusual connections.

There must be some even more fundamental difference (or differences) in dyslexic brains that accounts for both phonological processing problems and the other patterns of challenges and strengths associated with dyslexia. Next, and in chapter 4, we'll consider the remaining three dyslexia-related brain variations, each of which attempts to provide this more basic explanation.

Procedural Learning

The next key difference between dyslexic and nondyslexic brains to consider involves the *procedural learning system* and procedural memory.[4] One of the leading experts on procedural learning and dyslexia, British psychologist Dr. Angela Fawcett, described procedural learning and its relationship to dyslexia for us in the following way: "Procedural learning is learning *how* to do something, and learning it to the point where it's automatic, so you know how to do it without having to think about it. This process of becoming automatic with complex rules and procedures is much more difficult if you're dyslexic."

At least half the individuals with dyslexia have significant problems with procedural learning, and as a result they'll be slower to master any rule-based, procedural, or rote skill that should become automatic through practice. Because most basic academic skills are heavily rule and procedure dependent, problems with procedural learning can cause a wide range of academic challenges, which are often especially intense in the early grades.

For example, most language skills require the constant, rapid, and effortless application of rules and procedures, including differentiating one word sound from another; correctly articulating word sounds and correctly pronouncing words; breaking words down into component sounds; mastering the rules of phonics underlying reading (decoding) and spelling (encoding); recognizing rhymes; recognizing how changes in the forms of words can change word meanings and functions (morphology, e.g., *run*, *ran*, *running*, *runner*, *runny*, etc.); interpreting how differences in sentence organization and word order can affect sentence meaning (syntax); and recognizing language style and pragmatics (the language conventions that carry important social cues).

Many other academic skills are also rule based, such as rote (or automatic) memory of things like math facts, dates, titles, terms, or place names; memorizing complicated procedures or rules for things like long division, carrying over, borrowing, or dealing with fractions in math; sequences, like the alphabet, days of the week, or months of the year; writing conventions like punctuation and capitalization; and motor rules for forming letters the same way every time, when writing by hand, and spacing evenly between words.

Finally, individuals with procedural learning challenges also typically have difficulty learning simply by observing and imitating others as they perform the complete, complex skill—that is, by *implicit learning.* Instead, they learn better when rules and procedures are broken down into small, more easily mastered steps and demonstrated clearly—a process known as *explicit learning.* When you realize how important procedural learning is for most basic skills, you can see why procedural learning challenges have been thought capable of producing so many of the challenges associated with dyslexia.

Because individuals who struggle with procedural learning have difficulty learning to perform rule-based skills automatically, they must instead perform these skills using *conscious compensation*, or the combination of focused attention and active working memory. The drawback to this kind of highly focused processing is that if too many parts of a complex task must be performed consciously (because the basic skills haven't been mastered to the point where they're fully automatic), then working memory resources are very likely to be overwhelmed. Since individuals with procedural memory problems must perform many tasks using conscious processing, they will often experience working memory overload, which makes them slower and more error-prone than others on routine tasks.

Individuals with procedural learning challenges also tend to require many more repetitions than others to master complex skills. Dr. Fawcett explains:

"You can teach a dyslexic child what the rules are, and she appears to grasp them, but then the rules slip away again. We actually came up with something we called the *square root rule*, which means that it takes the square root *longer* to learn something if you're dyslexic than if you aren't. In other words, if it took four hours to learn something for a nondyslexic, it would take twice as long for a dyslexic; and if it took one hundred hours it would take ten times as long. So you can see how much extra work is needed to get these children to develop skills similar to other children."

Individuals with dyslexia and procedural learning challenges also tend to forget skills that they appear to have mastered more quickly than others if they don't practice them. "Often teachers will say, 'This child seemed to have learned this before the six weeks' holiday, but now he's come back and it's gone.' It helps to show teachers that it's not due to a moral fault in the student or any lack of effort, but it's really something to do with the basic learning processes. Actually, the dyslexic child is working much harder than everybody else, and this difficulty in learning and retaining rules results from a fundamental difference in the learning process. When you understand this, you realize that it's not something the child should be ashamed of, but something he should be taught to get around, using specific strategies."

From a neurological standpoint, procedural learning challenges are often associated with dysfunction in the cerebellum, a small, densely packed structure at the back and bottom of the brain. Although it accounts for only 10 to 15 percent of the brain's weight, the cerebellum contains nearly half the brain's impulse-conducting cells, or neurons. Although it was long thought to be involved primarily in helping with motor (or movement-based) functions, within the past decade scientists have come to realize that the cerebellum plays a critical role in most skills that become automatic through practice—whether those skills involve movement, language, "internal speech," working memory, or other aspects of attention.

There is now a wealth of evidence that at least half of all individuals with dyslexia experience difficulties with procedural learning. Typically, these individuals also show signs of mild cerebellar dysfunction on exam, such as low muscle tone; poor motor coordination; and difficulties with sequencing, timing and pacing, and time awareness.

This high incidence of procedural learning challenges in individuals with dyslexia has led Angela Fawcett and her collaborator, psychologist Roderick Nicolson, to propose the *procedural learning theory* of dyslexia, which posits that many of the findings of dyslexia are due to challenges with procedural learning. One of the great strengths of this theory is that it explains many of the symptoms commonly found in dyslexia that don't obviously relate to phonology or language, like challenges with motor control and coordination. We've found the procedural learning theory to be especially helpful for understanding and troubleshooting the learning challenges of individuals with dyslexia who show features like low processing speed scores on WISC IQ tests, very slow work output, motor problems with handwriting or eye movement control, problems with rote memory for things like math facts, more extensive problems with syntax or expressive language, special difficulties with sequencing, and poor time awareness and estimation.

Another strength of the procedural learning theory is that it predicts some of the advantages that we often observe in individuals with dyslexia. For example, while poor automaticity in routine skills makes many individuals with dyslexia slower and less efficient on routine tasks, it also forces them to approach these tasks with a greater "mindfulness" or task awareness and to really think about what they're doing. As a consequence, we've found that individuals with dyslexia often innovate and experiment with routine procedures, and in the process find new and better ways of doing things. In contrast, individuals with strong procedural learning abilities quickly learn to perform tasks in just the way they were taught, so they often perform these tasks without having to think about them. As a result, they less often feel the need to innovate. This kind of "flip side" benefit to dyslexic processing is just what we would expect to see from any full explanation of dyslexia.

Still, there are several drawbacks to the procedural learning theory as a complete explanation of dyslexia. Many individuals with dyslexia do not show clear procedural learning challenges, and many of the dyslexic advantages described in later chapters can't easily be attributed to increasing task mindfulness. For these reasons, a full explanation for both dyslexic challenges and the strengths must depend upon an even more fundamental feature of dyslexic brains. In the next chapter, we'll look at two variations in brain structure that may provide this deeper explanation.

CHAPTER 4

Differences in Brain Structure

In 1981, Dr. Roger Sperry was awarded the Nobel Prize for his discovery that the brain's two halves, or hemispheres, process information in very different ways. Ever since, a steady stream of books and articles have popularized the idea that there are distinctive "right-brain" and "left-brain" thinking styles and that individuals can be primarily "right-brained" or "left-brained" in their cognitive approach.[1] While these views of brain function are highly oversimplified, they still contain a good deal of truth: the brain's two hemispheres really do process information in very different ways.

As a rough generalization, the brain's left hemisphere specializes in fine-detail processing. It carefully examines the component pieces of objects and ideas, precisely characterizes them, and helps to distinguish them from each other. The right hemisphere specializes in processing the large-scale, big-picture, "coarse," or "global" features of objects or ideas. It's especially good at spotting connections that tie things together; at seeing distant similarities or relationships between objects or ideas; at perceiving how parts relate to wholes; at determining the essence, gist, or purpose of a thing or idea; and at identifying any background or context that might be relevant for understanding the objects under inspection.

We can roughly summarize the functional differences between left and right hemispheres by saying that they specialize respectively in trees and forest, fine and coarse features, text and context, or parts and wholes.[2] These differences show up in important ways in the brain's various processing systems. Consider vision: when looking at an object, the left hemisphere perceives fine details and component features, but it's poor at "binding" those features together to "see" the larger whole. For example, the left hemisphere can recognize eyes and ears and noses and mouths, but it's poor at recognizing faces. Similarly, it can see windows and doors and chimneys and shingles, but it's poor at seeing houses. To perceive these larger patterns, the left hemisphere requires big-picture processing help from the right.

We're raising this topic because several kinds of evidence suggest that individuals with dyslexia differ from nondyslexics in the ways they use their brain hemispheres to process information. In particular, a growing body of research suggests that individuals with dyslexia use their right hemispheres more extensively for many processing tasks than do nondyslexics. Differences of this type have been shown for many auditory, visual, and motor functions, and some of these differences play a role in reading and language.

This dyslexia-related difference in the division of labor between the brain's hemispheres is the third variation we'll examine in our search for the factors underlying dyslexic strengths and challenges.

Are Individuals with Dyslexia Really More "Right Brained" Than Nondyslexics?

Several prominent writers have observed that individuals with dyslexia often show a distinctly "right-brained style" or "flavor" in the ways that they process information. A particularly strong case for this connection has been made by author Thomas G. West in his marvelous book *In the Mind's Eye.*[3] West—who himself is dyslexic—suggests that this right-sided processing pattern may be directly related to the visual and spatial talents shown by many individuals with dyslexia.[4]

Scientists have also found that individuals with dyslexia use their right hemispheres more extensively for reading than do nondyslexics. This difference was first demonstrated in the late 1990s by Drs. Sally and Bennett Shaywitz at Yale, who used a brain scanning technique called functional magnetic resonance imaging (fMRI) to identify the brain areas that become active as individuals with dyslexia and nondyslexics read.[5] Reading expert Dr. Maryanne Wolf summed up the results of this work by writing, "The dyslexic brain consistently employs more right-hemisphere structures [for reading and its component processing activities] than left-hemisphere structures."[6]

While this increased right-hemispheric processing may at first appear to involve a "rightward shift" from the normal left-sided pattern, it actually reflects the *absence* of the usual "leftward shift" that occurs as individuals learn to read. Dr. Guinevere Eden and her colleagues at Georgetown University have shown that most beginning readers use *both* sides of their brain quite heavily—just like individuals with dyslexia. It's only with practice that most readers gradually shift to a largely left-sided processing circuit.[7]

Individuals with dyslexia have a much harder time making this shift to primarily left-sided, or "expert," processing. Without intensive training they tend to retain the "immature" or "beginner" pathway, with its heavy reliance on right-hemispheric processing.

This dyslexic tendency to retain the largely right-sided "beginner" pathway raises two important questions. First, *why* do individuals with dyslexia show this persistence of heavy right-hemisphere involvement? And second, what are the *consequences* of this persistence for dyslexic thinking and processing?

In approaching the first question, it's important to recognize that the reading circuit isn't the only brain pathway in which a right-to-left processing shift is produced by practice and experience. Transitions like this are seen in many brain systems, and they are thought to reflect our changing processing needs as our skills increase. The general idea goes like this.

When we attempt a new task, our right hemisphere's coarse or big-picture processing helps us recognize the overall point or essence of the task, so we don't get lost in the details. It also helps us recognize how the new task may be similar to tasks we've learned before, which helps us problem-solve and fill in details we miss. In these ways, the right hemisphere's top-down or big-picture processing is ideal for our early attempts to stumble through processes we're still fuzzy on. It's also invaluable when we try to tackle other tasks for which we lack the automatic skills to perform quickly and efficiently.

As we become more familiar with the purposes and demands of a task, our need for big-picture processing gives way to a demand for greater accuracy, efficiency, speed, and automaticity. That's where the left hemisphere comes in, with its greater ability to process the fine details that must be mastered to develop true expertise.

One well-documented example of a right-to-left-hemisphere processing shift that occurs with training is the shift that takes place as we develop musical expertise. Researchers have shown that untrained music listeners process melodies primarily with their right hemispheres, so they can grasp the large-scale features (or gist) of the melody. By contrast, expert musicians process music more heavily with their left hemispheres, because they focus on the fine details and technical aspects of the performance.[8]

This tendency to shift from right- to left-hemisphere processing as skill increases is intriguing because it suggests that the dyslexic failure to make such shifts might reflect a kind of general difficulty in acquiring expertise through practice. As we said in the last chapter, many individuals with dyslexia show precisely such a difficulty, especially in mastering rule-based skills like those involved in reading. Delays in mastering rule-based reading skills could clearly slow the development of "expert," left-sided pathways and cause prolonged dependence on the "novice," right-sided circuits. Similar difficulties in gaining expertise might also cause the greater right-hemisphere processing that individuals with dyslexia show with the other processing tasks we mentioned.

If delays in developing automaticity and expertise at least partly explain *why* individuals with dyslexia use their right hemispheres more for many tasks, then what are the consequences of this more right-hemispheric processing style? We can begin to answer this question by looking at differences in how the right and left hemispheres process information, using language as an example.

In 2005, Northwestern University psychologist Dr. Mark Beeman published a remarkable paper describing the differences in the ways that the two brain hemispheres process language. When the human brain is presented with a particular word, each hemisphere analyzes the word by activating its own "semantic field," or collection of definitions and examples describing that word.[9] Importantly, the semantic fields contained in the left and right hemispheres perform this analysis in significantly different ways.

The left hemisphere activates a relatively narrow field of information, which focuses on the "primary" (or most common, and often the most literal) meaning of the word. This narrow field of meaning is particularly well suited for processing language that's low in complexity or requires precise and rapid interpretation—like comprehending straightforward messages or following simple instructions. It's also useful for quickly and efficiently *producing* language. Since speaking and writing require the rapid production of *specific words* (rather than blended or compound words), the less ambiguity or hesitation the better. The left hemisphere's narrow semantic fields are ideal for such production.

The right hemisphere, by contrast, activates a much broader field of potential meanings. These meanings include "secondary" (or more distant) word definitions and relationships, like synonyms and antonyms, figurative meanings, humorous connections, ironic meanings, examples or cases of how the word can be used or what it represents, and words with similar "styles" (e.g., formal/informal, modern/archaic) or "themes" (e.g., relating to the beach, to chemistry, to emotions, to economics). This broader pattern of activation is slower, but it's also much richer. That's why it's particularly useful for interpreting messages that are ambiguous, complex, or figurative. Tasks for which the right hemisphere is particularly helpful include comprehending or producing metaphors, jokes, inferences, stories, social language, ambiguities, or inconsistencies.

We asked Dr. Beeman to illustrate the kind of "distant connection" that the right hemispheric semantic processing is particularly good at detecting. He responded with the following example. "Consider this sentence: 'Samantha was walking on the beach in bare feet, not knowing there was glass nearby.

Then she felt pain and called the lifeguard for help.' When most people hear that sentence, they infer that Samantha cut her foot. But notice that the sentence never explicitly states that she cut her foot, or even that she stepped on the glass. These facts have to be inferred, and these inferences are made by the right hemisphere. It produces these inferences by detecting the overlap in semantic fields between the terms *bare feet*, *glass*, and *pain*."[10]

The right hemisphere's special skill in making such distant and inferential connections is just what individuals with dyslexia need when reading and listening. Decoding problems often make it hard for individuals with dyslexia to identify printed words, and problems distinguishing closely related words can cause similar difficulties with listening. As a result, individuals with dyslexia must often use contextual clues to fill in parts of messages they've missed. This is just what the right hemisphere is so good at. Rather than causing reading or listening problems, the dyslexia-associated increase in reliance on right-hemisphere processing is actually an ideal compensation for individuals who are struggling to process language at its most basic levels.

The dyslexia-associated increase in right-hemisphere processing may help to explain many of the challenges and strengths that individuals with dyslexia commonly show. On the challenge side, the physically broader and more diffuse connections in the right hemisphere can lead to slower, less efficient, less accurate, and more effortful processing. This can place a greater burden on working memory, especially for tasks that require processing a great deal of fine detail. On the strength side, the broader network of connections provided by the right hemisphere favors new and creative connections, the recognition of more distant and unusual relationships, and skill in detecting inferences and ambiguities.

These are all promising points, but there is also a significant drawback to this theory: If it is true that the "dyslexic difference" in cognition is due entirely to the greater tendency to use right-hemispheric circuits, then we would expect that training sufficient to produce a right-to-left processing shift would cause individuals with dyslexia to become "just like everyone else" in their cognitive style. But this is not, in fact, what we see.

Consider, for example, the task of reading. Individuals with dyslexia who are trained sufficiently to produce the kind of right-to-left shift in their reading circuit that we described above usually don't become indistinguishable from fully "normal" readers but instead become their own unique variety of highly skilled "dyslexic readers." What we mean is that these dyslexic readers still generally read more slowly than comparably bright nondyslexics, and they also still display the same highly interconnected, gist- and context-dependent, imagery-based, big-picture reading comprehension style shown by most other individuals with dyslexia.

This persistence of big-picture processing despite the shift from right to left hemisphere suggests that there must be some even more fundamental factor underlying the processing style of dyslexics. A likely candidate for this "deeper" factor has only recently come to light, but already it appears to be the most promising candidate yet for the source of the dyslexic advantage.

Alterations in Microcircuitry: Big-Picture versus Fine-Detail Processing

This fourth and final dyslexic brain difference was first recognized by Dr. Manuel Casanova of the University of Kentucky School of Medicine. For the last two decades Dr. Casanova has studied the cell-to-cell connections that link the neurons—which are the cells most responsible for information processing—in the human brain. Given his broad interests as a psychiatrist, neurologist, and neuropathologist, Dr. Casanova has examined an enormous range of different "types" of brains, including those of clinically "normal" subjects and subjects diagnosed with a variety of cognitive or psychiatric conditions—including dyslexia.[11]

In analyzing the connections that link the neurons in the brain, Dr. Casanova identified one key feature that correlates both with a predisposition to dyslexia and with the kinds of "right-brain" cognitive style we've been discussing. This structural feature is an unusually broad spacing between the functional clusters of neurons in the brain's cortex. To explain why this discovery is potentially so important, we must first review a few things about the structure and function of the cortex.

The cortex is a thin sheet of cells that coats much of the brain's surface. The neurons in the cortex communicate with each other using a combination of chemical and electrical signals. In the process, they give rise to many of our higher cognitive functions, like memory, language, sensation, attention, and conscious awareness.

The cells in the cortex are organized into functional units called *minicolumns*: "columns" because they're vertically arranged, and "mini" because they're microscopic in size. Minicolumns were first discovered by researchers who inserted tiny electrodes into the brain's cortex to record its electrical activity. When they pushed their electrodes straight down into the cortex, like a birthday candle into a cake, they found that the cells stacked right on top of each other responded to stimuli in unison. In contrast, when they inserted the electrode at an angle to the brain's surface, those cells did not fire together. These results indicated that the cortical cells were grouped functionally into tiny columns that ran perpendicular to the surface of the brain: hence, minicolumns.

To process more than just the most basic kinds of information, minicolumns must be linked to form circuits, just as the microchips in your computer must be linked to create complex processing functions. Of course, unlike computer chips your minicolumns aren't soldered together. Instead, they're connected by long projections—axons—that extend like cables from the neurons in one minicolumn to connect with neurons in others.

When Dr. Casanova examined the arrangements of minicolumns and axons in many different brains, he found that each person showed a consistent pattern of spacing between his or her minicolumns. He also found that the degree of minicolumn spacing was distributed in the general population in a bell-shaped fashion, in which people on one end of the bell curve showed very closely packed minicolumns, while those on the other had very broadly spaced minicolumns.

Casanova also noticed that each individual's minicolumn spacing correlated closely with the size and length of the axons connecting the minicolumns in his or her brain: individuals with tightly spaced minicolumns sent out shorter axons that formed physically smaller or more local circuits, while persons with more widely spaced minicolumns sent out larger axons that formed physically longer-distance connections. In other words, individuals with tightly spaced minicolumns tended to form more connections between nearby minicolumns, while individuals with broadly spaced minicolumns tended to form more connections between minicolumns in distant parts of the brain.

This "bias" toward either long-range or local connections turns out to be highly important because there are enormous differences in how the circuits formed by these connections function and in the tasks at which they excel. Local connections are especially good at processing fine details—that is, at carefully sorting and distinguishing closely related things, whether different sounds, sights, or concepts. Brains biased to form more of these shorter connections generally show a high level of skill in detail-oriented tasks that involve "extracting" the fine features of objects or ideas.

In contrast, longer connections are generally weaker at fine-detail processing but excel at recognizing large features or concepts—that is, at big-picture tasks. Examples of big-picture tasks would include recognizing the overall form, context, or purpose of a thing or idea, synthesizing objects and ideas, perceiving relationships, and making unusual but insightful connections. Circuits formed from long connections are also useful for tasks that require problem solving—especially in new or changing circumstances—though they are slower, less efficient, and less reliable for familiar tasks and less skilled in discriminating fine details.

Notice how closely the strengths associated with short (or local) connections match the "left-brain" processing skills we discussed in the previous section; and see how closely the strengths associated with long (or distant) connections match "right-brain" processing skills. Also notice how closely the processing style associated with longer connections—that is, "strong big-picture/weak fine-detail"—matches the cognitive style we've described as being common among individuals with dyslexia.

Given these similarities, we might predict that "dyslexic brains" would tend to show a bias toward widely spaced minicolumns and physically longer brain circuits. And that is precisely what Dr. Casanova did find when he examined the connection patterns in the brains of dyslexic individuals.

We asked Dr. Casanova to explain in simple terms why a bias toward longer connections might favor big-picture processing. He responded that higher cognitive skills arise when minicolumns are connected to form a *modular system*. He illustrated what this would mean using the example of a car, which has many separate components, or "modules," such as the transmission, motor, and tires. When these modules are connected into a larger system, they can create new or *emergent properties* that aren't present in any of the separate modules—such as the property of locomotion. This example illustrates how in modular systems the properties of the whole can greatly exceed the properties of the individual elements and can create new functions that wouldn't exist if the parts were connected in some other way.

Dr. Casanova then explained how the same thing happens in the brain when minicolumns are joined into circuits. "Depending on how you link minicolumns to themselves, you get the emergence of higher cognitive functions, like judgment, intellect, memory, orientation. Those functions weren't there within the properties of the individual minicolumns. They emerged as the appropriate connections were made between cells in different parts of the brain. In other words, broader connections favor the formation of broadly integrated circuits, which in turn create high-level cognitive skills."

According to Dr. Casanova, the dyslexic bias toward long-distance connections leads both to the emergence of the big-picture processing skills we've mentioned and to weaknesses in fine-detail processing. One fine-detail task that Dr. Casanova cited as often being particularly hard for individuals with dyslexia is phonological processing, which, as we described in the last chapter, involves distinguishing highly similar sounds.

Difficulties with fine-detail processing could also explain many of the challenges with listening, vision, motor function, and attention we described in the last chapter. To further explain the characteristic dyslexic pattern of strengths and weaknesses, Dr. Casanova contrasted dyslexia with another well-known cognitive pattern.

"The brains of individuals with autism are biased toward short connections at the expense of long connections—just the opposite of dyslexia. Not surprisingly, when we looked we found a high proportion of individuals with autism in the other tail of minicolumn spacing, where the minicolumns are closely packed. Cognitively, individuals with autism focus on particular details: they see the trees, but lose the forest. If you test patients with autism, their thinking tends to be rather concrete, and they struggle to see the broader meaning, form, or context.[12] However, where they often excel is at tasks that can be performed using a tightly localized brain region, because they require only one specific function. An example would be finding Waldo in the *Where's Waldo* books. This fine-detail processing task is performed entirely within one highly localized area of the visual cortex, where the tightly packed minicolumns are connected by many short axons into a local circuit that excels in fine-detail processing. With such tasks, individuals with autism often perform much better than other people.

"On the other hand, where individuals with autism often struggle is with tasks like face recognition, which require that many different processing centers spread all around the brain work together. This joining or 'binding' of distant processing centers is very hard for individuals with autism, because they don't easily form the necessary long-distance connections.

"In contrast, joining distant areas of the brain together is just what individuals with dyslexia do best. As a result, individuals with dyslexia excel at drawing ideas from anything and anywhere, and at connecting different concepts together. Where they may miss the boat is in processing fine details."

The fact that this single variation in brain structure could predispose individuals to so many of the challenges and strengths that are associated with dyslexia strongly supports its potential importance.[13] It also strongly supports our idea that dyslexic processing isn't the result of a purposeless breakdown in function, but that it represents a valuable trade-off that's been chosen for its special processing benefits. The specific nature of these benefits will be our subject throughout the next four parts, on the MIND strengths.

Conclusion

In chapters 3 and 4 we've reviewed four dyslexia-associated variations in brain structure and function that we believe underlie many common dyslexic challenges and strengths. We've raised several key themes that we'll return to repeatedly as we go on to examine dyslexic strengths. The most important of these themes is that dyslexic brains are organized in very different ways from most nondyslexic brains because they're intended to work in different ways and to excel at different tasks.

Let's briefly review what we've learned about what excellent function means for typical nondyslexic brains versus dyslexic ones.

For many nondyslexic brains, excellent function consists of traits like precision, accuracy, efficiency, speed, automaticity, reliability, replicability, focus, concision, and detailed expertise.

For dyslexic brains, excellent function typically means traits like the ability to see the gist or essence of things or to spot the larger context behind a given situation or idea; multidimensionality of perspective; the ability to see new, unusual, or distant connections; inferential reasoning and ambiguity detection; the ability to recombine things in novel ways and a general inventiveness; and greater mindfulness and intentionality during tasks that others take for granted.

Nondyslexic brains often excel at applying rules and procedures in an expert and efficient manner. Dyslexic brains often excel at finding "best fits" or at ad hoc problem solving.

Nondyslexic brains often excel at finding primary meanings and correct answers. Dyslexic brains often excel at spotting interesting associations and relationships.

Nondyslexic brains often excel at spotting the differences and distinctions between things. Dyslexic brains often excel at recognizing the similarities.

Nondyslexic brains display the order, stability, and efficiency of train tracks, well-organized filing cabinets, sequential narratives, or logical chains of reasoning. Dyslexic brains store information like murals or stained glass, connect ideas like spiderwebs or hyperlinks, and move from one thought to another like ripples spreading over a pond.

In short, dyslexic brains function differently from nondyslexic ones not because they're defective but because they're organized to display different kinds of strengths. These strengths are achieved at the cost of relative weaknesses in certain kinds of fine-detail processing.

If you know anything about the conventional view of dyslexia, you know what the dyslexic mind looks like when it struggles with fine-detail tasks. In the following chapters, we'll show you what the dyslexic mind looks like when it opens its wings and begins to soar.

PART III

M-Strengths

Material Reasoning

CHAPTER 5

The "M" Strengths in MIND

You may not know Lance Heywood's name, but if you've ever caught a chairlift at a major American ski resort, ridden on a monorail anywhere between the Hilton Waikoloa Village in Hawaii and the Bronx Zoo in New York, or shuttled around Las Vegas on a people mover, you've probably encountered his work. Lance is one of the leading designers and producers of the electrical systems that control transportation units at entertainment venues all across the United States. It's a challenging job and one that requires constant innovation and on-the-spot problem solving. To do it as well as Lance you need a special kind of mind: creative, to meet the demands of different clients and their endlessly varied projects, and fully competent, to design products that are safe and reliable. Lance has these qualities in abundance, but there was little sign of this creative talent in his early schoolwork.

Lance grew up in what's now Silicon Valley, and from his earliest years he found reading and writing incredibly difficult. In fact, he needed constant tutoring all the way through middle school just to get by—even in math, where he eventually excelled.

In contrast, outside of school Lance found no shortage of fascinating things to captivate his mind. He was a "constant tinkerer" and especially loved working on projects with his father, who was an interior designer, gifted self-taught mechanic, and enthusiastic model-train hobbyist. Together they built radios, phones, and recording devices from kits and spare parts.

When Lance finally reached high school he began to enjoy the more challenging math and science classes that became available, and in those classes his grades improved. While he struggled to memorize formulas and equations, he found that by mastering the principles behind them he could generally derive them himself. And even though reading and writing remained tough, he found that by zeroing in on the ideas and opinions that his teachers thought were important he could earn decent grades.

Lance's strong performance in math and science eventually earned him admission to several competitive colleges, and he initially enrolled as a freshman at UCLA. However, Lance felt "lost" in a place that he found too enormous and impersonal, so he moved home and enrolled at Santa Clara University, where he thrived in the smaller environment. Lance shied away from courses that required much reading or writing, but through hard work and discipline he did well in his engineering courses and eventually earned his degree.

Lance then went to work for a contracting firm that designed electrical systems for high-rises in the San Francisco Bay Area. Although he enjoyed design work, he missed hands-on work with electronics. Hoping to combine his love of challenging projects with his love of skiing and the mountains, Lance went to work for a company that made ski lifts.

Eventually Lance grew tired of working for someone else, so in 1993 he set up shop on his own, and he's never regretted it. You can sense the enjoyment he still finds in his work when he describes how he tackles each project. He begins each new job from scratch rather than modifying previous projects, and he remains involved through every step of the manufacture, installation, testing, and fine-tuning of the electronic panels he creates.

Lance credits much of his design skill to his ability to mentally envision his projects in fully constructed form. As he reads his clients' proposals he envisions all the components he'll need coming together to form a three-dimensional blueprint in his mind, and he can manipulate these components at will. He told us that one of the things he most enjoys about his work is when he nears the completion of a project and he can finally see in the real world the creation he first envisioned in his mind.

Lance also credits his success to the fact that his slow reading and poor procedural memory always forced him to adopt a hands-on rather than book- or rule-focused approach. Although by age thirty Lance's reading had improved to the point where he could read for pleasure, he still reads slowly enough that he prefers to learn about new electronic parts and devices by interacting with them, rather than reading a manual or prospectus. As a result, he'll often find new uses for the equipment that are better than the task they were designed for.

We've shared Lance's story with you because, as you'll soon see, Lance is a perfect example of a dyslexic individual who excels in Material reasoning, the M-strengths in MIND.

Material Reasoning: A 3-D Advantage

> M-strengths are abilities that help us reason about the physical or material world—that is, about the shape, size, motion, position, or orientation in space of physical objects, and the ways those objects interact.

M-strengths consist primarily of abilities in areas that can be termed *spatial reasoning*, which has often been recognized as an area of special talent for many individuals with dyslexia. However, as we'll show in the next few chapters, dyslexic individuals with prominent M-strengths typically possess outstanding abilities in some areas of spatial reasoning but not others. In particular, the kind of spatial reasoning at which they excel involves the creation of a connected series of mental perspectives that are three-dimensional in nature—like a virtual 3-D environment in the mind.

This type of "real-world" spatial ability can be phenomenally valuable for the individuals who possess it. While M-strengths receive little emphasis or nurturing in most school curricula, they play an essential role in many adult occupations. Designers, mechanics, engineers, surgeons, radiologists, electricians, plumbers, carpenters, builders, skilled artisans, dentists, orthodontists, architects, chemists, physicists, astronomers, drivers of trucks, buses, and taxis, and computer specialists (especially in areas like networking, program and systems architecture, and graphics) all rely on M-strengths for much of what they do.

In the coming chapters, we'll look in detail at the nature and advantages of M-strengths and at the key mental processes underlying them.

CHAPTER 6

The Advantages of M-Strengths

Let's begin our examination of M-strengths by looking at two studies that compare the performance of individuals with and without dyslexia on various spatial tasks. In the first study British psychologist Elizabeth Attree and her colleagues compared dyslexic and nondyslexic adolescents on three different visual and spatial tasks. The first two tasks assessed two-dimensional spatial skills.[1] For the first task, the subjects were shown printed 2-D patterns, then they were asked to reproduce them using colored blocks. For the second task, the subjects were shown abstract line drawings for five seconds, then they were asked to draw them from memory. The third task was designed to assess the kind of three-dimensional spatial skills needed to perform well in a real-world spatial environment. For this task, the subjects were seated before a computer screen and asked to "search" through a virtual 3-D house to find a toy hidden in one of the rooms. After they'd searched the four "rooms," the computer was turned off and they were asked to reconstruct the house's floor plan from memory using cardboard shapes.

The dyslexic and nondyslexic groups performed very differently on the 2-D and 3-D tasks. While individuals with dyslexia did slightly worse than nondyslexics on the 2-D tasks (which stressed simple "snapshot" visual memory), they did much better on the virtual reality task, where they had to construct a seamless interconnected "world" from the views they'd absorbed during their explorations.

Notice how well this dyslexic strength/weakness profile fits our discussion in part 2: strength in the big-picture reasoning needed to combine multiple perspectives into a complex, global, interconnected, 3-D model of a virtual house, but relative weakness in fine-detail processing and memory. This is just the pattern of trade-offs we described.

These results have important implications for how we assess spatial ability on standardized testing. Many of the visual-spatial tasks commonly used on IQ and other cognitive batteries (such as block design and visual memory tasks) assess 2-D spatial abilities and fail to measure the kinds of real-world 3-D spatial strengths that individuals with dyslexia possess. When evaluating individuals with dyslexia for spatial skills, it's important to use tests that measure real-world 3-D reasoning skills.

A second study, this one by psychologist Catya von Károlyi, also supports the existence of a dyslexic 3-D/2-D trade-off.[2] Károlyi compared the abilities of dyslexic and nondyslexic high school students on two visual-spatial tasks. For the first task, subjects were asked to find an identical match for a complex 2-D Celtic knot pattern from among a group of four closely related patterns. This task required great accuracy in processing visual fine details. For the second task, subjects were asked to determine whether several drawings represented "potentially real" figures that could exist in 3-D space or "impossible" figures that couldn't. Success on this latter task required the ability to perceive how different parts of a figure related to each other in forming a larger whole—thus, big-picture or global (rather than fine-detail) processing.

The results of these studies perfectly mirrored Dr. Attree's in showing that dyslexic subjects displayed an advantage on tasks that required them to process multiperspective 3-D information, while showing a relative disadvantage on a "simpler" 2-D task. On the impossible figures task that stressed big-picture processing, Dr. Károlyi found that individuals with dyslexia answered significantly more quickly and just as accurately as the nondyslexic group. In contrast, on the Celtic knot task that stressed fine-detail processing, individuals with dyslexia were significantly less accurate than their nondyslexic peers.

At first glance, this skill in detecting impossible figures may seem rather far removed from real-world value, but we found an excellent example of its practical significance when speaking with a highly successful building contractor. This contractor, who is himself dyslexic, told us he prefers hiring dyslexic workers for his building crews because they excel at spotting flaws in blueprints that create "impossible figures" just like those in Dr. Károlyi's study. These kinds of 3-D spatial abilities are actually extremely valuable in many real-world occupations, and we can begin to understand the extent of this value when we examine how individuals with dyslexia put these skills to use at various stages throughout their lives.

The Real-World Worth of M-Strengths

Early in life, many dyslexic children with prominent M-strengths seem naturally drawn to engage in highly spatial tasks. In a survey of children from our practice (ages seven to fifteen), we found that children with dyslexia engaged in building projects—everything from LEGOs and K'NEX to small models to massive outdoor landscaping and construction projects—at nearly twice the rate of their nondyslexic peers. Even when these children engaged in 2-D art projects like drawing, their art tended to have a more multidimensional and dynamic quality, featuring elements like foreshortening and perspective, moving figures, arrows indicating action or process, and schematic elements like cutaway sections or multiangle or multiperspective blueprints. Dr. Jean Symmes, a research psychologist at the National Institutes of Health, similarly documented an unusually high interest and ability in building or visual classification tasks among the children with dyslexia she studied.[3]

While it's sometimes claimed that children with dyslexia gravitate toward high M-strength activities (and later occupations) because their reading and writing challenges make other activities too difficult, the preceding observations suggest that for most spatially talented dyslexic individuals, spatial interests and abilities are inborn rather than developed as compensations. Former Harvard neurologist Dr. Norman Geschwind—one of the most esteemed figures in the history of dyslexia research—noted that in his experience many dyslexic children display a passion and skill for spatial activities (like drawing, doing mechanical puzzles, or building models) well before they begin to struggle with reading.[4]

Another link between dyslexia and M-strengths is that children with dyslexia have parents who work in high M-strength occupations far more commonly than would be expected by chance.[5] In our own clinic we recently examined the employment and education histories of the parents of thirty dyslexic children. For twenty-two children, we were able to find one parent who either had personally shown dyslexic features or who had another first-degree relative (besides the child) with dyslexia; for another five children both parents showed such signs. Remarkably, nearly half of the thirty-two "dyslexia-linked" parents worked in high M-strengths jobs. This talented group included six engineers, three builders (construction, contracting, or development), two architects, two biochemists, two dental hygienists, and one inventor. The high frequency of engineers and architects among this group is particularly impressive. Together these two professions account for less than 6 percent of the college degrees awarded in the United States, but they accounted for 25 percent of the parents in our survey.

An even more direct link between dyslexia and high M-strength occupations is provided by several studies demonstrating that individuals with dyslexia are significantly overrepresented in training programs for highly spatial professions like art, design, and engineering. In the United Kingdom, where only about 4 percent of the population are considered severely dyslexic and another 6 percent moderately dyslexic, a study at the Royal College of Art found that fully 10 percent of its students showed severe dyslexic findings, and 25 percent at least moderate findings—*more than double the rates in the general population.*[6] At the Central Saint Martins College of Art and Design in London, psychologist Dr. Beverly Steffert found that more than 30 percent of the 360 students she tested showed evidence of dyslexia-related difficulties with either reading, spelling, or written syntax.[7] Another student survey at the Harper Adams University College in England showed that 26 percent of the first-year engineering students were significantly dyslexic—more than double the rate of the university's student body as a whole.[8] In Sweden, researchers Ulrika Wolff and Ingvar Lundberg compared the incidence of dyslexia in university students majoring in fine arts and photography with a control group of students studying economics and commercial law. They found that the art students showed nearly three times the incidence of dyslexia than among either the control students or the general population.[9]

While formal studies on the incidence of dyslexia among fully qualified professionals in these fields are lacking, there is no shortage of "occupational lore" in many high M-strength fields about the close connection between dyslexia and spatial ability. In his book *Thinking Like Einstein*, author Thomas G. West recounts his conversation with dyslexic computer graphic artist Valerie Delahaye, who specializes in creating computer graphic simulations for movies. Delahaye told him that at least half the graphic artists she's worked with on major projects like *Titanic* and *The Fifth Element* were also dyslexic. West also quotes MIT Media Lab founder and dyslexic Nicholas Negroponte as stating that dyslexia is so common at MIT that it's known locally as the "MIT disease."[10] Tufts University psychologist Dr. Maryanne Wolf has written that spelling difficulties are so widespread at the architectural firm where her brother-in-law works that they've instituted a rule that all architects must have their outgoing letters spell-checked—twice.[11] And author Lesley Jackson wrote in the design industry trade journal *Icon Magazine Online*, "Having met so many dyslexic designers over the years, I've become convinced there must be some kind of link between the underlying processes of design creativity and the workings of the dyslexic mind."[12]

Many dyslexia experts have also gone on record with their own observations regarding the links between spatial ability and dyslexia. Dr. Norman Geschwind wrote, "It has become increasingly clear in recent years that individuals with dyslexia themselves are frequently endowed with high talents in many areas. . . . There have been in recent years an increasing number of studies that have pointed out that many individuals with dyslexia have superior talents in certain areas of non-verbal skill, such as art, architecture, [and] engineering. . . ."[13] Eminent British neurologist Macdonald Critchley, who personally examined more than 1,300 patients with dyslexia, stated that "a great many" of these patients had shown special talents in spatial, mechanical, artistic, and manual pursuits, and that they frequently pursued occupations that made use of these abilities.[14] We could easily cite many more such observations.

The Cognitive Basis of M-Strengths

There are two key components to exceptional M-strengths. The first is an imagery system that can stably store and accurately display spatial information in a mental spatial matrix. The second is skill in manipulating these mental images by rotating, repositioning, moving, or modifying them, or by making them interact or combine with other mental images.

Recently, researchers at the University College of London have discovered a set of specialized cells in the brain's hippocampus (a complex structure at the base of the brain whose two seahorse-shaped lobes play many key roles in memory formation and spatial processing) that appear to be responsible for creating the brain's mental matrix, or 3-D spatial lattice.[15] They've named these cells "grid cells" because together they create a matrix of reference vectors that act like coordinate lines on a 3-D map.[16]

If it helps, you can picture these intersecting vectors as the bars of an infinite jungle gym. This spatial matrix allows us to plot information about where objects are in space—much like a 3-D GPS navigation system. This mental spatial coordinate system can help us interact with the real world, determining where we are in relation to other objects, or the sizes and shapes of those objects, or whether and how these objects are moving or changing in orientation. It can also help us reason about imaginary spatial environments or objects.

As we saw above, to be useful for real-world spatial reasoning, our spatial imagery must form a continuous 3-D web of interconnected perspectives. A simple "photographic snapshot"—no matter how vivid or detailed—is of limited use if it can't be manipulated or connected with other views and perspectives. The spatial coordinate system created by the grid cells helps—in cooperation with other functional centers of the brain—to tie these perspectives together.

This spatial information can be presented or "displayed" to the mind as various forms of *spatial imagery.* The most obvious form of spatial imagery is visual.

An excellent example of a dyslexic individual with impressive M-strengths and a remarkably clear and lifelike visual display of spatial imagery is Canadian entrepreneur Glenn Bailey. After academic problems caused him to drop out of school, Glenn became a highly successful businessman. One of his many successful ventures has been the development and construction of residential real estate. Glenn described for us how his ability to generate and voluntarily manipulate vivid, lifelike, 3-D visual imagery often helps him in this business. "When I see a property I can instantly construct a new house on it. I can see exactly how that house is going to look, and I can walk through every room in that house, and out into the garden, and everywhere. I can turn those thoughts into reality. And that's how my development company was created for high-end houses. Even right now, sitting here, I can do a detailed walkthrough in my mind of every house and property we've ever built."

Although stories like Glenn's might cause us to assume that strong visual imagery is essential for spatial reasoning, the experience of "MX" shows clearly that this assumption is false. MX was a retired building surveyor living in Scotland who'd always enjoyed a remarkably vivid and lifelike visual imagery system, or "mind's eye." Unfortunately, four days after undergoing a cardiac procedure MX awoke to discover that though his vision was normal, when he closed his eyes he could no longer voluntarily call to mind any visual images at all.[17]

MX was tested using a whole series of spatial reasoning and visual memory tasks. As a control, a group of high-visualizing architects performed the same tasks. Surprisingly, it was found that although MX could no longer create any mental visual images while performing these tasks, he scored just as well as the architects did. As he performed the tasks, MX's brain was also scanned with fMRI technology. In contrast to the architects, who heavily activated the visual centers of their brains while solving these tasks, MX used none of his brain's visual processing regions.

These studies suggested that while MX had lost his ability to *perceive visual images* when engaging in spatial reasoning, he could still *access spatial information* from his spatial database and apply it to Material reasoning tasks with no detectible loss of skill. In other words, MX had gone quite literally overnight from having remarkably vivid visual imagery to having none at all, without any apparent loss in his spatial imagery abilities. This is a dramatic demonstration of the difference between spatial reasoning and visual imagery.

Spatial imagery can actually be perceived in many ways besides clear, lifelike visual forms. As long as the hippocampus can create its spatial grid from information gathered through the senses, it seems relatively unimportant what form of imagery the individual uses to "read" or access this information. Think, for example, of a blind person who recalls the contours of a friend's face: this spatial information is recalled in a nonvisual form, as a form of tactile or "muscular" (*somatosensory*) imagery, yet it can be every bit as accurate and detailed as visual imagery.

We can also demonstrate the variety of useful spatial imagery styles by examining what other individuals with dyslexia with impressive M-strengths have said about their own forms of spatial imagery. Let's start with legendary physicist Albert Einstein.

In addition to having remarkable M-strengths, Einstein showed many dyslexia-related challenges, such as late-talking, difficulty learning to read, poor rote memory for math facts, and lifelong difficulty with spelling. Einstein described his own spatial imagery in the following way: "The words of the language, as they are written or spoken, do not seem to play any role in my mechanism of thought. The psychical entities which seem to serve as elements in thought are certain signs and more or less clear images which can be 'voluntarily' reproduced and combined."[18]

This kind of abstract imagery is especially common among spatially talented physicists and mathematicians, for whom the flexibility of such imagery seems to be particularly valuable. Dyslexic mathematician Kalvis Jansons, a professor at University College in London, has written, "To me, *abstract pictures and diagrams* feel more important than words. . . . Many of my original mathematical ideas began with some form of visualization."[19]

Jansons has also described experiencing spatial imagery in a completely nonvisual form—as feelings of movement, force, texture, shape, or other kinds of tactile or motor images: "It would be a mistake to believe . . . that non-verbal [spatial] reasoning has to involve pictures. For example, three-dimensional space can be equally well represented in what I often think of as a tactile world."[20] Jansons has employed this "tactile" spatial imagery in his professional work by using knots to study important principles of probability.

Dr. Matthew Schneps of the Harvard-Smithsonian Center for Astrophysics shared with us a related form of spatial imagery. Matt is an astrophysicist, an award-winning documentary filmmaker, and an individual with dyslexia. Matt described his spatial imagery to us as consisting of a feeling of movement or process—rather like a machine at work. When pursuing an idea or hypothesis, Matt sometimes feels like he's activating a lever in a machine he imagines in space, in order to turn a series of gears and observe how the spatial map changes as he tests various configurations.

Dyslexic attorney David Schoenbrod described to us still another type of nonvisual spatial imagery. David is Trustee Professor of Law at New York Law School, a pioneer in the field of environmental law, and a key litigator in some of the most important environmental cases of the last fifty years, including the landmark lawsuits that led to the removal of lead from gasoline. He is also a talented sculptor, architect, landscape designer, and builder. David described his spatial imagery to us as a strong sense of spatial position unaccompanied by clear visual images: "In recalling autobiographical stories I recall spatial arrangements in some detail—like the layout of the room, the arrangement of the furniture, where the other people and I were, and the ordination to the points of the compass—but this recollection is neither lifelike nor schematic, but rather in grayscale, and almost vanishingly faint. In general, I know the shape of things, but I don't really see them. It strikes me that the fact that I see form more than color is why I have been more attracted to sculpting than painting."

We've described these different forms of spatial imagery in some detail because we too often meet educators and individuals with dyslexia who believe that lifelike visual imagery is the key to spatial reasoning. As a result, they often overlook the value of other forms of spatial imagery. In truth, it doesn't seem to matter much whether your mental imagery is lifelike and visual or whether it is abstract, positional, or movement or touch related. So long as you can use this imagery to understand spatial relationships, you can use it to make important comparisons and predictions, or to combine, change, or manipulate spatial data in various ways. The ability to reason spatially is highly valuable for many tasks and professions, and individuals with dyslexia are often blessed with prominent M-strengths. However, as we'll discuss in the next chapter, M-strengths often come with several trade-offs.

CHAPTER 7

Trade-offs with M-Strengths

Two recurring themes in this book are, first, dyslexic advantages arise from variations in brain structure that have been selected for their benefit, and, second, these variations also bring "flip-side" trade-offs that can make certain tasks more difficult. As you'll see, each of the MIND strengths has its own set of trade-offs, and M-strengths are no exception. We've seen one trade-off already, and that's relative weakness in certain 2-D processing skills. While this weakness is of little consequence for most day-to-day functions, there's one area where it can create important problems: symbol reversals while reading or writing.

Struggles with Symbols

We often encounter two opposite and equally mistaken beliefs about symbol reversals and dyslexia. The first is that all young children who flip symbols are dyslexic. The second is that symbol reversing is never associated with dyslexia. To sort out the truth about this topic, we must examine how spatial skills develop in the human brain.

No child is born with the ability to identify the 2-D orientation of printed symbols—or of anything else, for that matter. The ability to distinguish an object from its mirror image is actually an acquired skill, and it must be learned through experience and practice.

Over the last decade researchers have found that the newborn human brain forms two mirror-image views of everything it sees: one in the left hemisphere and the other in the right. Usually this duplicate imagery is helpful because it allows us to recognize objects from multiple perspectives, so that a toddler who's been warned about a dog while looking at its left profile can recognize that same dog from its right.

Unfortunately, when trying to recognize the orientation of printed symbols—or any other item with a natural mirror, like a shoe or glove—this ability to generate mirror images becomes a burden. Before a child can reliably distinguish an image from its mirror, he or she must *learn* to suppress the generation of its mirror image.[1]

Some children have an especially hard time learning to suppress this mirroring function. When they first learn to write, many children will reverse not only symbols that have true mirrors (like *p/q* or *b/d*), but essentially all letters or numbers. For most children these mistakes begin to diminish after only a few repetitions. However, until the age of eight as many as one-third of children continue to make occasional mirror image substitutions when reading or writing. If such mistakes are only occasional and the child has no real difficulty with reading and spelling, these errors are neither important nor a sign of dyslexia.

However, for some truly dyslexic children—in our experience roughly one in four—letter reversals can be a much more persistent and important problem. These children may reverse whole words or even whole sentences, and at the single symbol level they may reverse not only "horizontal" mirrors like *b/d* or *p/q*, but also "vertical" mirrors like *b/p*, *b/q*, *d/p*, *d/q*, or *6/9*. They may make so many reversals when reading that it worsens their comprehension.

Published studies have shown that many younger dyslexic children have more difficulty rapidly determining letter orientation than their nondyslexic peers, though this difficulty declines with age.[2] In our experience persistent reversals—not just for letters and numbers, but even for drawings and other visual figures—are most often a problem for those who are most gifted with M-strengths. Leonardo da Vinci is an extreme example of this. His lifelong dyslexic difficulties in reading, word usage, syntax, and spelling were combined with phenomenal M-strengths. While many people are aware that Leonardo wrote his journals in mirror-image script, few know that he also drew many of his sketches and landscapes in mirror image.

We've worked with many individuals with dyslexia who've continued to reverse symbols when reading—or more commonly writing—well into their college years and beyond. Most of these individuals experience sporadic errors, but we met one student who unintentionally "lapsed" into writing full paragraphs in mirror image whenever she grew tired. Probably not coincidentally, she's now a graduate student in architectural history.

One reason that spatially talented individuals with dyslexia may be especially susceptible to reversals is that their brains are just so good at rotating spatial images. Listen to dyslexic designer Sebastian Bergne: "If I'm designing an object, I know the exact shape in 3-D. I can walk around it in my head before drawing it. I can also imagine a different solution to the same problem."[3] While this image flexibility may be useful when you're trying to design a chair or a teapot, it's less useful when you're trying to read or write symbols on a 2-D surface. Dyslexic biochemist Dr. Roy Daniels was one of the youngest members ever elected to the prestigious National Academy of Sciences, but even as an adult he still confuses mirrored letter pairs like *b/d* and *p/q* both when reading and when writing. To compensate, he does all his handwriting in capital letters, "to help me tell the difference between letters like *b* and *d*."[4] Dr. Daniels is far from unique in this regard.

It's likely that difficulties with procedural learning, which we discussed in chapter 3, may contribute to these persistent reversals because the ability to turn off the symmetrical image generator is itself a kind of procedure that must be learned through practice.[5] As a result, it will be mastered more slowly by individuals with dyslexia who show procedural learning challenges.[6]

Ease of Language Output

A second trade-off that we often see in individuals with dyslexia with prominent M-strengths is difficulty with language output. Parents and teachers are often puzzled to find that their bright dyslexic students struggle to answer apparently "simple" questions—especially in writing. This difficulty can be particularly intense when the questions are open-ended and students are given a great deal of latitude in how they respond. Difficulty answering questions of this kind is one of the most common reasons why older dyslexic students are brought to our clinic. We've found that this difficulty is often particularly bothersome for dyslexic individuals with high or even gifted-level verbal IQs, because the ideas these students are attempting to express are often so complex.

The research literature suggests several possible reasons why dyslexic individuals with impressive M-strengths may be especially vulnerable to expressive difficulties. First, some of the brain variations associated with dyslexia may enhance spatial abilities at the direct expense of verbal skills. Psychologists George Hynd and Jeffrey Gilger have described one such variation. In this structural variation, brain regions that are usually used to process word sounds and other language functions[7] are essentially "borrowed" and connected instead to brain centers that process spatial information. Drs. Hynd and Gilger first identified this brain variation in a large family with many members who showed both dyslexia and high spatial abilities. They then identified this same variation in the brain of Albert Einstein, who, as we've mentioned, displayed a similar combination of spatial talent and dyslexia-related language challenges.

Einstein's comments on his own difficulties putting his ideas into words provide useful insight into the challenges many of our high M-strength dyslexic individuals experience. Although Einstein eventually became a talented writer, he once complained that thinking in words was not natural for him, and that his usual mode of thinking was nonverbal. To communicate verbally he needed first to "translate" his almost entirely nonverbal thoughts into words. Einstein described the process this way: "[C]ombinatory play [with nonverbal symbols] seems to be the essential feature in productive thought—before there is any connection with logical construction in words or other kinds of signs which can be communicated to others. . . . Conventional words or other signs have to be *sought for laboriously* [italics added] only in a secondary stage."[8]

We've found that many individuals with dyslexia—and especially those with prominent M-strengths—identify closely with Einstein's descriptions both of his primarily nonverbal thinking style and of his difficulties in translating his thoughts into words. While translating nonverbal thoughts into words can be difficult at any stage of life, it is often especially difficult for children and adolescents, whose working memory capacities are still far from fully developed. This is likely one reason why children from families with a high degree of spatial and nonverbal attainment are often slower than other children to begin speaking.[9]

In fact many (though not all) high M-strength individuals with dyslexia reason in largely nonverbal ways and often find it difficult to translate their thoughts into words. This means that they will often show a gap between their conceptual understanding and their ability to express or demonstrate that understanding in words. It's important that those who work with these individuals be sensitive to this challenge. There's a long and quite shameful tradition among certain psychologists and educators of treating "nonverbal reasoning" as if it were at best a poor cousin of verbal reasoning and at worst a kind of oxymoron—like "civil war" or "act naturally."

In fact, nonverbal reasoning is real, scientifically demonstrable, and often a key component of creative insights of all kinds, and it deserves to be taken seriously in all its forms. While students with dyslexia should try their best to express their thoughts in words, it's also critical that parents, teachers, and later employers learn to recognize that some valid forms of reasoning may be difficult to put into words and may be better expressed as drawings, diagrams, or other forms of nonverbal representation.

Besides this fairly direct trade-off between spatial and verbal ability, studies have also demonstrated a more indirect way that strong spatial and visual imagery skill can hinder verbal functions. Dr. Alison Bacon and her colleagues at the University of Plymouth in England asked dyslexic and nondyslexic college students to supply a valid conclusion to a series of syllogisms for which they'd been given major and minor premises.[10] For example, if they were given the premises "All dogs are mammals" and "Some dogs have fleas," they were asked to provide a conclusion, such as, "Some mammals have fleas."

The researchers found that the dyslexic students reasoned just as well as their nondyslexic peers when they were given premises that provoked little imagery (e.g., all *a* are *b*, no *b* are *c*, how many *a* are *c*?), or when the visual imagery contributed directly to the solution of the syllogism (e.g., some *shapes* are *circles*, all *circles* are *red*, how many *shapes* are *red*?). However, when the syllogisms contained terms that provoked strong visual imagery *unrelated to the reasoning process* (e.g., "Some snowboarders are jugglers, all horsewomen are snowboarders, how many horsewomen are jugglers?"), the dyslexic students performed significantly *worse* than the nondyslexics. The authors concluded that their vivid mental imagery was swamping their working memory and hindering their verbal reasoning.

This potentially distracting role of visual imagery has important implications for how we teach dyslexic students with strong imagery abilities. Think, for example, how needlessly burdened a student with strong imagery abilities will be by visually elaborate story problems in math. Many teachers have been taught that using imagery helps children with strong spatial and visual skills, but this is true only if the imagery is directly useful for solving a problem. Irrelevant imagery is distracting and worsens performance.

A final point to remember about language development in individuals with dyslexia—and especially those with prominent M-strengths—is that their language is simply developing along a different pathway than that followed by their nondyslexic peers. The brain systems that help "translate" nonverbal ideas into words are some of the latest-developing parts of the brain. For many children and adolescents with dyslexia, difficulty putting complex ideas into words is a normal feature of development and one that diminishes with maturity. That's why their progress must be judged by its own standards, rather than by standards that apply to the nondyslexic population. Focusing too much on their challenges can make us overlook their special strengths, as we observed with one very special child, Max.

CHAPTER 8

M-Strengths in Action

As an infant Max was late to start talking, and when he finally began to speak it was in a language all his own: *ma* was water, *dung gung* was vacuum cleaner, and *wow wow* was pacifier. When he started preschool at age three and a half, Max's mother remembered, he had difficulty "catching on to things that the other kids seemed to simply absorb. He never learned the songs or rhymes, couldn't remember the names of the other kids, and could rarely retell what happened during the day." In first grade he went to a Montessori school, but "he didn't 'discover' academic knowledge and skills on his own. He needed to be explicitly taught."

Through the end of first grade Max made little progress in reading, math, or writing. He seemed to have a hard time staying focused. He also struggled to retrieve words and information from memory. His kindergarten and first-grade teachers found that he seemed to learn much better one-on-one than in a large class, so Max's mother decided to homeschool him for second grade.

One-on-one, Max slowly began to learn. Although he still required frequent repetitions and refocusing, by the end of second grade he was reading—slowly—and his writing began to take off (though it largely left his spelling behind). The following is a response he wrote to the question "Tell me about going to the [Seattle] Science Center":

> we went a long wae and thin we wint in sid. And we qplab [played] with the ecsuvatr [excavator] and thin we trid too pla with the tic tac toe mushen [machine] and thin we wint too the bug thing and thin we wint too the binusho [dinosaur] thing and thin we wint toe the ecsuvatr and thin we left.

Outside of school, Max found plenty to occupy his time. When he was a toddler he became fascinated by wires and circuits, and as he grew older he developed a sophisticated interest in electronics. He especially liked experimenting with small-scale power generation, using sources like solar, wind, and water to generate electricity.

Max also developed a deep interest in nature, and he loved to spend time in the woods surrounding his home near Seattle. Max laid out a nature trail on his family's property, but as many Northwesterners have learned, he soon found that "walking" plus "woods" equals "wet feet." So Max began to install an elaborate series of drains to remove the standing water that collected across his trail. He also built bridges over the spots he couldn't drain. Max's drainage project was remarkable for a child not yet ten years old. Not surprisingly, it did draw remarks—from psychologists.

When Max was in fourth grade his mother took him to be evaluated for his difficulties with schoolwork. The psychologist diagnosed ADHD. This seemed plausible given Max's problems with auditory-verbal working memory, distractibility, and lack of focus for schoolwork. However, the psychologist was also concerned by Max's "intense" interest in electronics and drainage, his focus on solitary pursuits, and his difficulty talking with—and like—other children. So in addition to ADHD, the psychologist diagnosed Max with Asperger's syndrome, an autism spectrum disorder. As one of his recommendations, the psychologist suggested that Max take social skills classes.

Although Max's mother questioned the Asperger's diagnosis, she agreed that Max needed to improve his social skills, so she took him to see a speech-language pathologist (SLP). Fortunately, the SLP understood that social skills consist largely of complex rules for behavior that have been learned and practiced until they've become habits. Under the SLP's supervision, Max was taught social skills in a clear and explicit manner, and he practiced them during structured interactions with another child until these skills became automatic.

Max continued to improve both socially and academically. Between ages seven and a half and ten his reading vocabulary shot up from the 35th to the 98th percentile, and his math calculation rose from the 45th to the 99.9th. However, he still showed the lower Working Memory and Processing Speed scores that we typically find in our "young engineers."

At that point, Max was making great strides in most areas, though his reading comprehension and fluency lagged behind his conceptual abilities, and his writing remained slow. He also made frequent errors in his writing with spelling, conventions, sentence structure (syntax), and organization. The following is an essay that Max wrote at age ten, which he entitled, "The Derte road."

> This trip we whent to bary my Grate Grandma on this dert road. So when we got on the road in are fourrunner my teeth were chattring but they stoped when we got on the bumpy port. It was 13 miles in to the drte road so I just relaxed . . .

We met Max shortly before his eleventh birthday. While the "numbers" from our testing mostly confirmed what others had found, our interpretation was somewhat different. We identified his challenges with reading and writing fluency, spelling, syntax, rote and working memory, focused attention for auditory-verbal material, processing speed, sequencing, and organization, but we also found many of the wonderful strengths that individuals with dyslexia often show. Max showed tremendous spatial and nonverbal reasoning powers, his understanding of math concepts was amazing, and his ability to interact knowledgeably on a wide range of complex scientific subjects was extremely impressive. We were also struck by the intellectual "flavor" Max displayed as he approached his work. He showed a charming naïveté and inventiveness on many tasks that children his age usually dash through without even giving a thought, so although his work was slower it was often more creative. Max also made many interesting observations that showed how his mind was reaching out to probe the connections between ideas.

We also discovered several important details about Max's family. Max's father has a Ph.D. in chemistry, and his mother has a degree in biochemistry—both high M-strength fields. Max's mother also has dyslexia-related processing traits and talents. Although she struggled with reading and writing as a child, she now works as a medical writer. Both of Max's maternal grandparents were also scientists, and Max's great-grandmother has often remarked how much Max reminds her of his grandfather when he was a boy. Perhaps "family resemblance" provides a better explanation for Max's "intense" interest in drainage and erosion than autism, since his grandfather spent his career as a professor of geophysics at UCLA.

Ultimately, we made several suggestions to help Max in areas where he still struggled, but we also explained that in the most important respects Max was right on target for *his* development—that is, for the kind of late-blooming growth and maturation that children display when they combine dyslexic processing, outstanding M-strengths, and procedural learning challenges.

M-Strengths and Development

We've shared Max's story because we want you to see what dyslexic children with impressive M-strengths look like while they're still developing—before their mature talents are fully apparent. While it's enormously helpful to look at the childhoods of successful dyslexic adults, sometimes the perspective provided by hindsight can make their successes seem almost inevitable—as if their challenges weren't really so severe, and they were never really at risk of failure. Somehow these stories can lack the power to convince us that the slow, awkward, inarticulate, inattentive, and dyslexic early elementary child before us might actually have a chance to become one of the great engineers, architects, designers, mechanics, inventors, surgeons, or builders of the twenty-first century. Yet this is often unquestionably true.

We constantly meet skeptics who respond by pointing out that not every child will become an Albert Einstein or an Isaac Newton. This is true, but even Einstein and Newton didn't *look* like "Einstein" and "Newton" in second grade: Einstein was remembered as a slow, uncooperative child with a nasty temper who repeated everything he said (echolalia), while Newton was remembered as a simpleton whose only apparent use was to make small wooden toys for his sisters and schoolfellows.

Even though evidence of classroom success may be thin for many high M-strength children with dyslexia, they often display their creative potential quite clearly outside the classroom, in their desires to build, experiment, draw, and create. Recall compact disc inventor James Russell building his remote-control boat at age six, Lance Heywood "tinkering" on his electronics projects, or Max building his nature trail and experimenting with solar power. These activities should be taken much more seriously because for a dyslexic child with substantial M-strengths, a toy is never "just a toy" or a drawing "just a doodle." These activities provide a window into their future, and failure to regard them seriously does these children a grave disservice. One amazing young child with dyslexia whom we saw in our clinic built a K'nex structure so elaborate that it won second prize in a nationwide competition; yet when he brought it to school and asked his teacher if he could show it to the class he was told, "We don't have time for that; we have important work to do." Another was scolded by his teacher for doodling: "If you spend all day drawing buildings on your papers, you'll never get anywhere." Ironically, this child's father is a successful architect who makes his living in just that way.

When he was young, pioneering neurosurgeon Dr. Fred Epstein showed a sustained interest only in mechanical activities, like building elaborate model airplanes. Because of his dyslexia, Epstein barely made it through college and was initially rejected by all twelve of the medical schools to which he applied. However, as an adult Epstein developed many new and highly innovative surgical techniques for treating previously inoperable spinal cord tumors—techniques that saved literally thousands of children's lives. It's important to realize that when Epstein was devising these techniques he wasn't using skills he'd picked up in some classroom but the skills he'd developed at his workbench in the garage, building model airplanes.[1]

To identify our next generation of talented engineers, inventors, and physicists, we shouldn't be using pencil-and-paper "talent searches" or seeing who's fastest at the "mad math minutes." We should be searching for spatial prodigies in LEGO Stores and hobby shops—just like athletic scouts hang around ball fields. Many of our next generation's great Material reasoners are currently struggling in school while their talents are going unrecognized, and we owe it to them to pay closer attention to the ways that they typically develop.

CHAPTER 9

Key Points about M-Strengths

Material reasoning—the ability to reason about the physical characteristics of objects and the material universe—represents one of the most common and important talent sets found in individuals with dyslexic processing styles. Key points to remember about M-strengths are:

- The ultimate purpose of M-strengths is to create a continuous, interconnected series of 3-D perspectives as a basis for reasoning about real-world, global, or big-picture spatial features, rather than about fine-detail or 2-D features.
- The spatial imagery perceived by individuals with M-strengths may take many forms, from clear visual imagery to nonvisual perceptions like force, shape, texture, or movement.
- The form that spatial imagery takes is less important than the uses to which the reasoner can put it.
- M-strengths often bring trade-offs like symbol reversals and subtle language challenges.
- Individuals with dyslexia in general—and those with prominent M-strengths in particular—show a late-blooming pattern of development, and their developmental progress should be judged on its own terms, rather than by standards created to judge nondyslexics.

- Individuals with dyslexia who show prominent M-strengths may struggle in the early grades but often show signs of impressive creativity outside the classroom.
- Dyslexic children with prominent M-strengths have tremendous potential and often grow up to become remarkable and creative people.

Let's end this section by returning full circle to where we began: with Lance Heywood and his family. Not long after we interviewed Lance, we got a call from his wife, Jenny. She had some questions about their older son, Daniel. We'd seen Daniel in our clinic several years earlier. Like his father, Daniel is dyslexic; and like his father, Daniel has remarkable M-strengths.

For the last five years, Daniel has supplemented his homeschooling curriculum with courses at a state university. His coursework has focused on the spatially related disciplines in which he excels, like physics and higher math. Like his father, Daniel often solves problems in his own unique ways, rather than using procedures he's been shown in class; but he usually gets the answers right, as his outstanding grades attest.

Daniel's also been a member of the university's robotics team, and for the last several years he's traveled with them to compete against other universities. One year he designed a crucial component for the team's Mars rover, which finished second in a national competition.

Jenny informed us that Daniel was now applying for full-time enrollment at several colleges with strong engineering and biorobotics programs. Daniel wants to learn to build medical devices for people with physical disabilities. Jenny mentioned one of the schools to which Daniel was applying, and we told her we had a contact there Daniel might like to meet—the department chair, in fact.

We'd seen her children. For dyslexia.

PART IV

I-Strengths

Interconnected Reasoning

CHAPTER 10

The "I" Strengths in MIND

"Everything is about relationships. Things are as they are because of their relationships with everything else. You can't just look at anything in isolation."

As Jack Laws spoke to us by phone from his home in San Francisco, it became clear that his view of relationships and interconnection wasn't just a throwaway line but a true expression of his way of understanding and experiencing the world. It's a view he's shared with the growing number of readers of his amazing field guides on the wildlife of California, published under the name his parents gave him in tribute to another great California naturalist: John Muir Laws.

Anyone who's read Jack's field guides or attended his lectures knows that he's a teacher of exceptional skill. But as a child Jack never expected he'd have anything to teach because he found it so hard to learn.

Jack was in the second or third grade when he first realized "there was something odd going on with my brain." Even though he was clearly bright and hardworking, he seemed unable to learn many things his classmates mastered easily, like reading, getting his letters to face the right way, and memorizing math facts.

Jack's parents took him for evaluation, and he was given a diagnosis of dyslexia. Targeted learning therapy greatly reduced—but didn't eliminate—his challenges with reading, spelling, writing, and math. "The therapists tried to tell my teachers how to help me learn, but the diagnosis of 'dyslexia' wasn't a part of their training, and they weren't in a place where they could really hear it. This was new stuff, and I had teachers who thought that the best way to make me learn was to embarrass me more in front of the rest of the students—and that didn't really work for me. So I was bumped around to a number of different schools. I was smart enough to see that other kids could do this stuff, but try as I might I couldn't, and over time I became increasingly convinced that I just wasn't one of the 'smart kids.'"

While Jack got little encouragement in the classroom, outside of school there were several "bright rays of light" that buoyed his spirit and encouraged the growth of his mind. "My dad was an amateur bird-watcher, and my mom was an amateur botanist; so on all of our family trips we would study nature together. I kept field notes of my observations, but it was difficult for me to write; so instead I made lots of diagrams and sketches of the birds that I would see, and the flowers I would find, and what the bugs were doing, and those sorts of things. Those notebooks became my way of training myself to look more closely at the world around me. If I wanted to sketch an object, I had to look at it again, and again, and again, and I discovered that even the most common things still held secrets and discoveries for me."

It was on these family outings that Jack first recognized that an intricate web of relationships connected everything in nature. "On these family trips, I was either down on my belly looking at flowers with Mom, or finding birds with Dad, or exploring around the mountain meadows catching frogs. So I wasn't just out in flowers, or out in birds: I was out in a place that had all of these different elements, and they were interacting with each other. So my sketchbooks were full of whatever I was looking at—not only bugs or only birds, but from a very early start, I was actually looking at ecosystems."

Jack's other "bright ray" was Scouting. "I had a wonderful experience in Boy Scouting. I became an Eagle Scout and a leader in my troop, and I discovered that I could tie the fastest bowline knot in the San Francisco Bay Area Council. I was really good at leading a group of kids . . . solving problems . . . helping people get along with each other . . . demonstrating first-aid skills. And I really did well with map reading: I could look at a topographical map and the landform would three-dimensionally pop out at me, and I could route-find better than anybody else. That helped me have more self-confidence, which was a really important part of my childhood. But academically . . . school: not so much."

It wasn't until his sophomore year of high school that Jack finally began to find his way as a student. Appropriately enough his breakthrough came from connections he made with two of his teachers, one in biology and the other in history. Jack credits the profound impact they exerted on his life to the fact that they "stopped looking at my spelling and started looking at the content of what I was writing. There was a revolution in my brain because they helped me see that I had good ideas—I just couldn't spell them right. And when I finally realized that those were two different things, that was *huge* for me. It still took me forever to finish things, but for these two teachers I would have done anything. Between the two of them, in one semester they turned me around: they just opened up the door, and I went through it and started running. It was tremendously exciting. My life, and the turns that it has taken, is very much due to their influence."

While Jack considered careers in both history and biology, he realized, "If I went the history route I'd have to read a lot more and write a lot more, but with biology I could run around outside and listen to birds."

Jack's choice of biology—and his life's vision for his work—was finally (and fittingly) crystallized on a school trip when his class hiked the John Muir Trail. "On that trip I started fantasizing about a perfect field guide that would have full-color pictures of everything I was seeing."

For years the production of that field guide would remain Jack's cherished goal and the subject of countless daydreams. "I could visualize whole sections of the book and what the pages would look like and how I would organize it."

The next few years were a time of tremendous growth for Jack. "At the end of high school and going into college I was getting the idea that there were different things I could do to compensate for the things I struggled with. Once you realize that you've got something really good inside you to share, then dyslexia becomes a much smaller obstacle, and you learn to compensate and deal with it." Recorded books and a word processor with spell-check were especially helpful in allowing Jack to succeed in school. "Not being afraid of technology and embracing what can help you is crucial. I still can't spell and I still don't know my multiplication tables, and I've never read a book from cover to cover without books on tape, but hey—it's okay! I'm doing just fine." "Just fine" for Jack includes a bachelor's degree in conservation and resource studies from the University of California–Berkeley, a master's in wildlife biology from the University of Montana, and a degree in scientific illustration from the University of California–Santa Cruz.

After completing this training, "I picked up my backpack and headed up to the Sierra Nevada and started painting wildflowers and animals. I didn't stop for the next six years."

We mentioned to Jack that the marvelous empathy for his subjects that he conveys in his field guides reminds us of the remarkable ability to "get inside" the minds of animals that we've seen many children with dyslexia display. "It's funny you should say that," he responded. "My connection with nature is not just an intellectual one. It's deeply spiritual, and empathic. . . . When I give lectures about natural history, all these animals in my head are not just balls of factoids about this species or that species, but they're characters, they have personalities, I give them voices. Some people are very careful not to anthropomorphize the species that they see—and of course it's not scientifically correct to project a human perspective on all the things around you; but while I recognize that my human perspective is different than the world that's perceived by a wolverine, I still find myself constantly speaking for the wolverine or the pika and essentially trying to put myself in their shoes and give commentary about things from their perspective.

"I sometimes get this vision: I'll sit in a place, and I'll lean my back against a tree, and I'll look at something, and I'll think, 'How does that relate to something else that's here?' Then I'll imagine a sort of colored line of energy going between those two things. Then I'll look at another thing and see how that relates to it, and I start trying to actually picture in my head the web in front of me of relationships between things: a three-dimensional web. And I start thinking to myself, 'This is just some of the stuff that I've experienced or studied about,' and there are so many more that I don't know about or understand.

"John Muir said when you try to pick out any one thing by itself, you find it hitched to everything else in the universe. And the more we dig around as scientists and biologists, the more we see that that is absolutely true. You can't just look at anything in isolation. I don't know if seeing connections is easier for me, but sometimes I can actually visualize them. I just imagine the dense web of networks floating in front of me, and it's very clear to me that I'm a part of that web, as well: those lines are connecting to me, and I'm connecting to those lines."

Interconnected Reasoning: A Web of Meaning

Jack Laws is an outstanding example of an individual with impressive I-strengths, or Interconnected reasoning. Individuals with prominent I-strengths often have unique ways of looking at things. They tend to be highly creative, perceptive, interdisciplinary, and recombinatory. No matter what they see or hear, it always reminds them of something else. One idea leads on to the next. We often hear from their parents, teachers, spouses, and colleagues things like, "They just see connections that other people miss," or "She'll often say things that seem so off topic, then five minutes later I'll finally get it and see that she jumped right to the key issue."

I-strengths create exceptional abilities to spot connections between different objects, concepts, or points of view. They include:

- The ability to see how phenomena (like objects, ideas, events, or experiences) are related to each other, either by "likeness" (similarity) or "togetherness" (that is, association, like correlation or cause and effect).

- The ability to see phenomena from multiple perspectives, using approaches and techniques borrowed from many disciplines.
- The ability to unite all kinds of information about a particular object of thought into a single global or big-picture view and to determine its gist, or most essential or relevant aspects in particular contexts.

You may wonder as you view this list how individuals with dyslexia could excel at making complex connections like these when they struggle to form connections that most people make easily—like the connections between sounds and symbols, or between basic math equations and their answers. The answer to this apparent paradox is easy: not all connections are alike.

In part 1, we saw that M-strengths help to create a 3-D spatial matrix that can be used to understand and manipulate spatial information, and that this spatial matrix is especially useful for understanding global or big-picture spatial relationships. In this part, we'll see that I-strengths also work to create a multidimensional matrix, but this matrix is *conceptual* rather than spatial. Like the spatial matrix, this conceptual matrix aids in organizing and manipulating information, and it appears to reflect dyslexic strength in forming large-scale brain circuits that are especially suitable for processing big-picture connections rather than fine details. In the next chapter we'll examine the three kinds of I-strengths that form the "dimensions" of this conceptual matrix.

CHAPTER 11

The Advantages of I-Strengths

The power of Interconnected reasoning lies in its ability to link all of an individual's knowledge, ideas, and mental approaches into an integrated conceptual matrix. This integrated matrix is incredibly powerful because it allows objects of thought to be approached from many different angles, levels, and perspectives, so they can be seen in new ways, related to other phenomena, and understood in a larger context. The three core skills, or I-strengths, that help form this conceptual matrix are the abilities to detect relationships between different objects of thought, the ability to shift perspectives or approaches, and the ability to reason using a global or top-down perspective.

Strength in Perceiving Relationships

The first I-strength is the ability to detect relationships between phenomena like objects, ideas, events, or experience. These relationships are of two types: relationships of "likeness" (similarity) and "togetherness" (or association, such as correlation or cause and effect).

Relationships of *likeness* can link physical objects, ideas, concepts, sensations, emotions, or information of any kind, and the likenesses may range from highly literal to purely figurative. Some individuals with dyslexia excel at detecting only particular types of relationships (for example, those linking visual patterns or those linking verbal concepts), while others show more general strengths.

Several published research studies support the idea that individuals with dyslexia, as a group, show special talents for finding similarities and likenesses. In a paper published in 1999, English psychologists John Everatt, Ian Smythe, and Beverly Steffert compared the performances of dyslexic college students and their nondyslexic classmates on two tests of visual-spatial creativity.[1] Both tests measured the ability to recognize similarities between different objects or shapes—that is, to see how one thing could represent or replace another. The first was an "alternative uses" task, in which the students were asked to name as many uses as they could think of either for soda cans or for bricks. The second was a "picture production" task, in which the students were asked to form as many different pictures as they could using five different geometric shapes.

The results were striking. On both tasks the students with dyslexia handily outperformed their nondyslexic peers, imagining 30 percent more possibilities on the alternative uses task and drawing over one-third more pictures on the picture production task. Everatt later reported similar results when he tested younger individuals with dyslexia.[2]

We've noticed that many of the individuals with dyslexia we see in our clinic show remarkable strengths in detecting similarities among objects, structures, or physical patterns. Many are amazingly skilled at recognizing the works of particular designers, artists, or architects by their characteristic style, often before they're old enough to read. Many are also prolific inventors, builders, or sculptors. They often show special skill in adapting whatever materials they find at hand to construct their projects, demonstrating an unusual ability to see analogies between these "spare parts" and other objects. On our Dyslexic Advantage website (http://dyslexicadvantage.com) you can see an outstanding example of this ability in the work of Mariel, a talented student with dyslexia who—like Jack Laws—is a gifted naturalist as well as an artist. When she was eleven, Mariel won a highly competitive regional art competition with one of her "junk sculptures" made from bits and pieces of discarded objects.

Many individuals with dyslexia are also highly skilled in detecting similarities between words and verbal concepts. These connections can include analogies, metaphors, paradoxes, alternate word or concept meanings (especially more distant or "secondary" meanings), sound-based similarities (like homonyms, rhymes, alliterations, or "rhythmically" similar words), and similarities in attributes or categories (like physical appearance, size, weight, composition, and uses and functions).

We first noticed that individuals with dyslexia often excel in spotting verbal or conceptual connections during our testing sessions, when we observed differences in how individuals with dyslexia understood and employed words and verbal concepts. As we discussed in chapter 4, when processing a word or concept, many individuals with dyslexia activate an unusually broad "field" of possible meanings, rather than a tightly focused one. As a result, they are less likely to respond first with the primary or most common answers, and more likely to give unusual or creative answers, or a range of possible meanings and relationships.

For certain tasks, this perception of broader conceptual linkages can create an advantage. For example, individuals with dyslexia often show exceptional strength in spotting associations in tasks that require them to connect items conceptually. In one task that's a part of a common IQ test, examinees are given three sets of pictures, then asked to find one picture from each set that can be linked by a common concept. This task can unleash astonishing creativity among dyslexic examinees. Though we've been using the same pictures for many years, we still receive entirely new answers to these questions from our dyslexic examinees.

Dyslexic examinees are also more likely to detect more distant and unusual connections in verbal similarities tasks. For example, when we ask how *blue* and *gray* are similar, most people respond with the "obvious" answer, that both are colors. In contrast, dyslexic examinees sometimes give answers like, "They're the colors of the uniforms in the Civil War," or "They're the colors of the ocean on sunny and stormy days." Recently, a very bright seven-year-old boy with dyslexia responded to our question "How are a cat and a mouse alike?" by proudly announcing, "They're the two main characters on a highly popular children's television program called *Tom and Jerry*."

We also frequently observe this heightened ability to identify distant and unusual connections when we ask individuals with dyslexia to interpret ambiguous sentences—that is, sentences that can correctly be interpreted in more than one way. Ambiguities in word meanings are common in English because many words have multiple meanings or may be used as different parts of speech (noun, verb, adverb, etc.). In fact, the five hundred most commonly used words in the English language have an average of twenty-three different meanings *each* in the *Oxford English Dictionary*.[3] Correctly identifying which meaning is intended in a particular sentence requires a processing system that's capable both of identifying many possible meanings and of choosing the appropriate meaning based on the overall context of the sentence. During our testing, we commonly ask examinees to find two or more meanings for sentences that are intentionally ambiguous, such as:

The chickens are too hot to eat.
I saw her duck.
Please wait for the hostess to be seated.
Students hate annoying professors.
They hit the man with the cane.
I said I would see you on Tuesday.

We often find that dyslexic students who've struggled with many other fine-detail language tasks are able to correctly interpret these sentences with no difficulty at all, while students who've excelled at many of the fine-detail language tasks may struggle. Skill in recognizing alternate meanings is useful for interpreting all kinds of complex messages, like stories, jokes, conversations (especially informal ones), poems, and figurative language of all kinds (like analogies or metaphors). As we discussed in chapter 4, this skill is also highly useful for reading, especially for struggling readers.

Individuals with dyslexia who possess prominent I-strengths also frequently show an impressive ability to spot relationships of *togetherness*—that is, correlation or cause and effect—between things, ideas, or experiences. This ability is sometimes referred to as *pattern detection*, and pattern detection has often been noted by experts on dyslexia to be a special dyslexic strength.[4] Strength in detecting relationships of correlation or cause and effect is a useful skill in many fields, including science, business, economics, investment, design, psychology, leadership, and human relationships of all kinds. Jack Laws expressed his awareness of the pervasiveness of cause-and-effect relationships in nature in his description of the "three-dimensional web [of] relationships between things" that he often sees, and in his recognition that "things are as they are because of their relationships with everything else."

Another dyslexic scientist who has demonstrated an acute perception of the interconnectedness of nature is Dr. James Lovelock. Lovelock is best known as the formulator of the Gaia hypothesis, which states that the climatic and chemical components of the earth's crust and atmosphere interact to form a complex system that maintains the earth in "a comfortable state for life."[5] Lovelock was first led to posit such connections when he noticed subtle correlations in the variations of the chemical composition of earth's atmosphere and oceans. While other scientists before Lovelock had recognized that the earth's atmosphere was almost perfectly suited for biological life, none had realized that this special balance was maintained by the interactions of a tightly linked network of chemical processes: they'd observed the same parts but had missed the interconnections that form the whole system.

Strength in Shifting Perspectives

The second I-strength is the ability to see connections between different perspectives, approaches, or points of view. This I-strength helps its possessors see that a particular problem, idea, or phenomenon can be studied using different approaches and techniques, borrowed from different disciplines or professions.

Individuals with this I-strength prefer interdisciplinary rather than specialized approaches when taking on new problems or projects. They typically reject traditional ways of categorizing knowledge into "watertight" fields or disciplines and are dissatisfied with narrow and highly reductive approaches. Instead, they try to use as many different approaches as they can to solve problems and further their understanding, and they often borrow and adapt techniques from many different sources, applying them in new ways.

This way of viewing information often leads individuals with dyslexia to become "multiple specialists" who are knowledgeable in several fields rather than highly specialized in a single one. As a result of this interdisciplinary mind-set, they often find new and creative ways to apply approaches from one field to others where they're not usually used. Sometimes their recognition that interesting questions require new approaches leads them to seek broader training. For example, James Lovelock already had a Ph.D. in physiology when his growing interests in environmental science and climatology led him to pursue a second doctorate in biophysics. Ultimately, it was this blending of professional perspectives that suggested to him that the earth's biosphere might be understood and studied as if it were a physiological system.

In the past, many individuals with especially severe dyslexic challenges, unable to complete their professional training through traditional academic routes, were forced to acquire their skills through work experience or self-education. For example, John "Jack" Horner always knew he wanted to become a paleontologist, but after dyslexic difficulties caused him to flunk out of college seven times, he was forced to abandon conventional schooling. Eventually Horner worked his way up from an entry-level post as a museum technician and became one of the world's foremost paleontologists. Today he's both a full professor at Montana State University and curator of paleontology at the Museum of the Rockies.

The special combination of strengths and challenges that many individuals with dyslexia possess often causes them to have unusually varied life histories and careers. We've often thought that if the life stories of many of these individuals were placed on a series of "story tiles," it would be impossible to assemble the tiles in the correct sequence. However, in retrospect, the twists and turns their lives take usually make sense—they just couldn't be predicted in advance. The logic of Interconnected reasoning that directs their life journeys is so completely different from traditional logic that unless it's viewed carefully it may seem to be no logic at all.

Strength in Global Thinking

The third and final I-strength is the ability to combine different types of information into a single unified big-picture or global view. This I-strength reflects the ability to perceive the "whole" that can be made by combining different "parts" and to identify its central essence. This critical skill is one of the chief components of the dyslexic advantage, and as we discussed in chapter 4, it results from a fundamental variation in the way that dyslexic brains are organized and structured to form larger circuits, which unite multiple processing centers. We can better understand this big-picture reasoning strength by examining one of its key components: *gist.*

Gist is the main point, essence, or overall meaning of a thing, idea, concept, or experience. It's the "rough," coarse, or bird's-eye view, rather than the fine-detail view: the forest rather than the trees. Gist deals with overall shape or contour, or the core meaning. Gist detection helps us recognize the overall context or setting, so we stand a better chance of being able to fill in any information we find ambiguous or unclear and to determine what's relevant and what's not. For example, when a telephone or radio message is garbled by static, we use gist detection to determine the context of the message as a whole, then we use this knowledge of context to fill in the details we've missed.

All verbal messages have a gist, but this gist can't be determined by simply "adding up" all the primary meanings of each of the words in the message, then "computing" the global meaning like a sum. Instead, the gist or overall meaning of the message must be distilled by carefully considering all the possible meanings of all the words and phrases, then determining the essence of the message as a whole. Through this search for gist, clues about the source's meaning and intent can be identified, as can the source's mood and style.

These clues reveal whether the source is trustworthy, flippant, alarmed, emphatic, etc. Ultimately, these gist-determining skills lie very close to the core of what we mean by "understanding," and they're essential for determining the meanings of all but the simplest verbal messages and especially for complex messages like stories, plays, poems, jokes, or social interactions.

Gist also lends power to the other I-strengths. Gist detection allows us to determine the fitness of analogies and metaphors. It also helps us decide which perspectives, viewpoints, or approaches we should use to best understand some object of thought. In these ways, gist can be thought of as the ability to detect the relevant context or broader background of an object, idea, or message.

Researchers have shown that individuals with dyslexia tend to rely more heavily on gist detection than do nondyslexics for comprehending verbal information. T. R. Miles and colleagues at the University of Bangor in Wales presented both dyslexic and nondyslexic college students with four sentences of increasing complexity and measured how long it took each student to master the verbatim repetition of the sentences. While none of the twenty-four nondyslexics needed over eight repetitions to master the fine details, the dyslexic students averaged considerably more trials and some required as many as twenty-five. Despite these difficulties in mastering the fine details, the dyslexic students performed just as well as the nondyslexics in retaining the gist of the sentences.[6]

Many dyslexic readers are particularly dependent upon background clues and context to home in on gist. That's why many dyslexic readers show better comprehension for longer rather than shorter passages—*especially if the extra passages contain helpful contextual clues*. We often listen as dyslexic students misread every second or third word of a complex passage and wonder how they'll ever understand what they're reading. Yet when the passages provide helpful background context, we often find that the comprehension of even severely slow and inaccurate readers is surprisingly strong and in some cases outstanding. This kind of "upside surprise" is due largely to their ability to use contextual clues to grasp the gist of the passage, which allows them to correctly guess at the identities of the individual words.[7] In contrast, when passages contain few contextual clues, comprehension usually worsens as the passages grow longer.

This beneficial effect of extra context is why many dyslexic readers enjoy books that are part of a series. Series books contain many of the same characters, settings, and activities and often use similar words. A similar improvement in comprehension can be seen when individuals with dyslexia are pre-equipped with a summary of the passage they'll be reading, shown a film version, or supplied with a list of key words.

This skill in gist detection is also very helpful in areas besides reading. Many individuals with dyslexia and prominent I-strengths develop a settled habit of searching carefully for gist and context in all areas of life. As a result, they often look for deeper and deeper layers of meaning and context beneath the obvious meanings. This pattern of continually "peeling back the onion" to find the deeper significance of an idea or thing or occupation is one we've seen repeatedly in the individuals with dyslexia we've studied. Often it plays a key role in their success, leading them to question things and ideas that have long been taken for granted, and allowing them to find secrets that have been "hidden in plain sight," in a way that often makes others ask, "Why didn't I notice that?" We'll illustrate the value of this gist-detection skill in a later chapter, when we discuss the work of a highly successful individual with dyslexia.

CHAPTER 12

Trade-offs with I-Strengths

Just like the rest of the MIND strengths, I-strengths are often accompanied by cognitive trade-offs. In fact, each of the I-strengths we've discussed so far comes with its own set of "flip-side" challenges.

Trade-offs with Strength in Perceiving Relationships

While the ability to see broad fields of meaning is useful for detecting relationships of similarity, or "likeness," the tendency to identify more "distant" or secondary relationships rather than to fix immediately on primary connections can also worsen performance in certain settings. It's especially likely to cause problems in settings where speed, accuracy, reliability, and precision are more valued than creativity, novelty, or insight.

One such setting is standardized tests, including the IQ tests we described before. For example, in tasks like linking picture concepts or identifying verbal similarities (which we described in the last chapter), many individuals with dyslexia are so extravagantly good at coming up with insightful but nonprimary connections that we often find their test scores—which are based entirely on the number of "right" or primary answers they give—misleading.

This dyslexic talent for finding unusual connections can also lead to difficulties in the classroom. Most of the tasks students are asked to perform in school—like reading fairly literal texts, responding to simple questions, or acting on straightforward instructions—are easier for minds that routinely fix on primary meanings. Students who call up more distant meanings can appear confused or "off target," especially if they don't "get" the simple answers that everyone else does or if they become confused by ambiguities that no one else detects.

This broader pattern of associations can also worsen speed, precision, accuracy, and reliability on tasks that are best approached in a straightforward and literal matter. A classic example is the multiple-choice exam. Cynics might be forgiven for suspecting that multiple-choice exams—with their terse, dense, noncontextual sentences—were designed specifically to trip up individuals with dyslexia who excel in detecting secondary meanings or distant word relationships. These examinees will often pore over a multiple-choice exam like a lawyer vetting a contract, finding loopholes, ambiguities, and potential exceptions where none are intended. While their classmates evaluate questions with a "reasonable doubt" standard, they search for "proof beyond the shadow of a doubt." As a result, even a hint of uncertainty leads them to reject answers that most students would identify as correct. If their reading is also somewhat dodgy, the multiple-choice exam usually becomes a nightmare. But even dyslexic students who read longer passages or whole books with excellent comprehension may struggle with multiple-choice exams.

For some individuals with dyslexia, each word or concept may be surrounded by such a rich network of associations that these associations can become overwhelming and give rise to unintended substitutions. Sometimes these substitutions involve "near-miss" or similar-sounding words,[1] like *adverse/averse*, *anecdote/antidote*, *persecute/prosecute*, *conscious/conscience*, *interred/interned*, *imminent/eminent*, *emulate/immolate*. While such errors are usually attributed to problems with sound processing (that is, difficulty with phonological awareness), a careful examination of dyslexic word substitutions suggests that factors other than impaired word sound processing are often also involved. Consider Mark, whom we saw in our clinic. Mark is a highly creative boy with a great fund of knowledge and a lively imagination. Yet he often struggles to say what he means. Sometimes his verbal substitutions involve words with similar sound structures, such as:

"There were three people out in the missile." [middle]
"I was looking at an add column." [ant colony]
"Look at the winnows." [minnows]
"Those people are cocoa." [cuckoo]
"Being dizzy can really affect your carnation." [coordination]
"That purple light caused an obstacle illusion." [optical]

At other times, though, Mark substitutes words that bear only a slight structural similarity to the intended word (e.g., sounds, length, "rhythm") yet share some relationship of meaning:

"We made this for Dad's graduation." [celebration]
"Max, quit ignoring me." [annoying]
"Jim was there at the book club." [chess club]

At still other times, Mark substitutes words with almost no structural similarity, so that the relationships are purely conceptual:

"Don't eat that—it will spoil your breakfast." [dinner]
"That was a great Valentine's, wasn't it!" [Christmas]
"Mom, where's the bacon?" [baloney]
"Those curtains have polka dots." [stripes]

Conceptual substitutions like these are referred to as *paralexic* or *paraphasic* errors; and when made during reading they're sometimes called *deep substitutions.* While they are less common than sound-based errors, in our experience more individuals with dyslexia make them (at least occasionally) than is generally supposed.

In her autobiography, *Reversals*, dyslexic author Eileen Simpson vividly describes her frequent paralexic substitutions. One example she cites is her unintentional substitution of the word *leaf* for *feather*—a conceptual rather than sound-based substitution. Over time, Simpson learned to cover such slips (when they were pointed out to her) by pretending that they were intentional puns or jokes.[2]

We believe this tendency to substitute related items is the flip side of dyslexic strengths in perceiving distant conceptual relationships. In support of this idea, we've found that the individuals who make these substitutions the most often excel on tests that require the ability to spot distant connections, like ambiguities or similarities.

Special strength in recognizing relationships of "togetherness" also comes at a cost. That's because skill at detecting correlations or causal relationships has been shown to be enhanced if your attention system is a little bit distractible. Many studies have found that individuals with dyslexia experience difficulty screening out irrelevant environmental stimuli, like noises, movements, visual patterns, or other sensations. This sensitivity to environmental distractions is one of the main reasons dyslexic students often need special accommodations for testing and other work that takes focused concentration: they're just not that good at automatically screening out irrelevant environmental stimuli. These distractions invade their conscious awareness and steal working memory resources.

At the other end of the distractibility scale, the ability to quickly and subconsciously distinguish relevant and irrelevant stimuli so you can ignore those that are irrelevant is called *latent inhibition*. Latent inhibition sounds like an unmixed blessing, and it's definitely useful in circumstances requiring tight attentional focus—like tests or silent work time at school. In fact, latent inhibition makes you the kind of student most teachers dream of having. However, before you conclude that students who test high in latent inhibition and low in distractibility are the lucky ones, you should know that there's an *inverse* correlation between latent inhibition (or freedom from distraction) and creativity. What this means is that the highest creative achievers tend to score lower on tests of latent inhibition and to be somewhat distractible. In fact, one study looking at Harvard students showed that nearly 90 percent of those who showed unusually high creative achievement scored *below average* in latent inhibition—just like individuals with dyslexia.[3] This is a critical fact to keep in mind when evaluating the balance between focus and distractibility in individuals with dyslexia.

Trade-offs with Strength in Shifting Perspectives

The second I-strength—the ability to shift between perspectives—is also remarkably useful, so long as you recognize when the shifts are taking place and they're under your control. However, we often find that individuals who can shift perspectives easily are subject, especially when younger, to switching perspectives without realizing it, and this can complicate certain tasks. For example, on a biology paper dealing with animal behavior, a student may begin by describing behaviors, then shift (appropriately) to a discussion of the neurological sources of that behavior, then veer off topic to consider other points of neuroscience that don't relate to the central topic. The student may also bring in elements of personal experience or opinion where they don't really belong and forget that he's writing a scientific treatise rather than an autobiography or opinion piece. Often such students' papers will have an air of free association that can be fascinating but takes them far from where they need to go. For students with strong perspective-shifting abilities, learning to control the team of horses that's tied to their mental chariot often takes great effort and prolonged and explicit training.

This ability to shift perspectives and to see things in interdisciplinary ways can create problems with organization as well. For example, high I-strength individuals with dyslexia are often horrific filers of papers. This is not, as is often supposed, simply because they have trouble alphabetizing but because they can think of too many places to file each paper and are more likely to lose papers that have been filed neatly away in distinct folders. As an alternative, they frequently prefer to keep papers in stacks where they can more easily find them. There are some wonderful pictures of Einstein's office at Princeton that beautifully illustrate this dyslexic "filing system." (They can be found online by searching "Einstein's office.") Fortunately, hyperlinked computer files and search capabilities have helped to reduce this problem.

Trade-offs with Strength in Global Thinking

The primary trade-off with strength in global or big-picture thinking is that it can create a greater dependence upon context and background information. Global thinkers have a top-down reasoning style that works best when a big-picture overview is already in place, so that new chunks of information can be added to conceptual frameworks that have already been built. That's why big-picture thinkers often learn best when they have at least a general understanding of the goals or ends at which they're aiming.

Big-picture, top-down learners are often a poor fit for the typical classroom, where bottom-up teaching approaches predominate. Schools often ask students to memorize new bits of information before explaining their meaning or significance. This approach doesn't really work for top-down learners because they can remember only things that make sense to them and new information that can be related to other things they already know. If they can't see the point of something they're asked to learn, it just won't stick. Without a big-picture framework to hang their knowledge on, the information is simply incomprehensible.

This dependence upon context is why "stripping down" instruction to the bare minimum to avoid overloading individuals with dyslexia often results in failure. Individuals with dyslexia who have a top-down, big-picture learning style typically learn better from approaches that convey information with greater conceptual depth, rather than from more superficial or survey-type approaches.

Individuals with this style also show several other characteristic patterns. For example, it's common to find dyslexic global thinkers at the upper levels of schooling still feeling lost far into the term, then suddenly finding that things become clear when enough of the big pieces are finally available to reveal the whole picture. Students with this pattern are often also more aware of (and bothered by) the gaps and deficiencies in the things they've been taught because they're more aware that parts of the picture are still missing. Students with this pattern typically do better the longer they stay in school: upper-division college courses are generally easier for them than entry-level ones, and graduate school and postdoctoral work go even better than college.

For individuals with dyslexia who have this highly interconnected learning and conceptual style, a few simple steps can help them learn more effectively and enjoyably. For longer reading assignments, providing them with an overview (gist and context) of assigned passages beforehand can improve their reading speed, accuracy, and comprehension. If any new or special vocabulary will be included, giving them a list of key words in advance can be very helpful. Previewing the practical relevance and applicability of the information they'll be asked to master will improve retention and motivation. Tying in new information with things they've already learned also improves memory and comprehension. And beginning each new course or unit by previewing the major points that will be asserted, and the route that will be taken to demonstrate them, can keep dyslexic students better oriented, more confident, and better able to learn.

To demonstrate how I-strengths and the challenges that go with them can appear at various stages of development, let's look in the next chapter at an individual with dyslexia who excels in Interconnected reasoning. His name is Douglas Merrill.

CHAPTER 13

I-Strengths in Action

When Douglas Merrill was young, he struggled to make basic academic connections. As he told us, "Reading was—and is—challenging, so getting through assignments meant using a bunch of tricks." With writing, "every other letter was backward." And with math: "Every summer my mother was reteaching me to add, subtract, multiply, and divide all the way up till I was in college. . . . Math never clicked for me. Even when I was in high school I failed algebra."

Not surprisingly, Douglas's self-esteem suffered: "I always felt defective, which caused the sorts of things you'd expect in a kid, like superdefensiveness and hostility, because I felt like I was failing."

Douglas labored to get by on extra effort and sheer force of will, but the results were disappointing. It was only when he reached middle school that he realized that despite his difficulties with rote and fine-detail tasks, he also had special strengths. One of his most important strengths was his ability to think and communicate using stories. "Pretty early on, I started writing stories to answer problems instead of doing what the assignment actually asked. So if you look at my junior high school papers, what I was doing for most of my classes was writing short stories. I was never going to be able to remember all the details that would be required to lay out a terse step-by-step outline. But I could remember the story arcs."

Douglas gradually realized how widely this technique could be used, but he was slowest to realize its relevance for math. "The breakthrough came in high school when I failed algebra. My dad's a Ph.D. in physics, and my brother's a Ph.D. in math, and one of my sisters is a practicing physicist working in nuclear power plants, so everyone but me is great in traditional math. When I failed, it forced me to look for something I did unusually well because I had to find some way to balance how awful I felt. I realized I could tell overall stories better than most people, and when I looked through what I had done wrong in algebra, I realized I was playing exactly to my weaknesses. I was not trying to make the math into a story. I was trying to memorize 'step A, step B, step C' by rote without creating any meaningful story about how they were connected, and that maximized my likelihood of failing."[1]

Though school remained hard, Douglas began to make slow but steady progress. After graduating from high school he went on to the University of Tulsa, where he discovered a special fondness for studying the big ideas and forces that shape our world, and for interdisciplinary approaches to those topics: "I have always been interested in the overlap between psychology, sociology, and history; the three work together to constrain what we can do, how we can do it, and how we view ourselves."[2] Unable to limit his focus to one subject, Douglas dual-majored in economics and sociology. In his spare time—which he claims to have had plenty of, as a self-described "can't-get-a-date geek"—Douglas picked up another skill set that eventually played a big role in his life: he became an expert at cracking computer programs and in computer security.

After college, Douglas combined his interests in big-picture questions, learning and cognition, and computing by pursuing a Ph.D. in cognitive science at Princeton, where he performed groundbreaking research on learning, decision making, and artificial intelligence. When we asked him what motivated him to study cognitive psychology, Douglas responded, "Pure vindictiveness. I'd spent a lot of my life thinking through tips and tricks and different ways to solve problems, and I thought it would be interesting to think through how you could model that problem solving—how to think about it and how to formally describe it."

As he studied these questions, Douglas realized that his insights into his own thinking could be useful for others, too. "When I investigated human problem solving, one of the things I studied was people learning math and programming—partially out of a sense of irony—and I found that even 'normal' people who are pretty good at the traditional rote skills do better if they form stories, with starting and plot elements and end of a story, for even things like writing programs. Now, no one I studied did this to the extent that I did, but I demonstrated a huge problem-solving improvement when you teach 'normies' to do what I did instinctively. That's something traditional schools could definitely teach to make their students more successful."

After earning his doctorate, Douglas went to work as an information scientist for the RAND Corporation, a prestigious think tank that researches public policy questions. At RAND, Douglas combined his expertise in decision making and computer security to perform studies for clients—including the U.S. government—on computer security and "information warfare," or attempts by hostile entities to cripple an information system. Douglas found that his ability to think in stories and to see "the big picture" was incredibly useful in helping him imagine methods of attack that could exploit those weaknesses and to detect gaps in information security systems. Douglas also found that at RAND his big-picture, interdisciplinary, story-based thinking style was a perfect fit. As he told us, "RAND . . . is fundamentally a narrative place, so you're telling policy stories about what ought to happen, backed up by data, and that plays really well to my strengths. I can work with the people who do the data analysis itself, and I can say, 'Oh, here's where we're going.' I didn't understand that at the time, but that skill of being able to say, 'You know, I think we're going to head over there . . . ,' was actually super useful—especially when I left to go into business."

Douglas's unique combination of skills eventually brought him to the attention of private-sector recruiters and to a series of high-level jobs at Price Waterhouse, Charles Schwab, and eventually a small Bay Area start-up with the unlikely name of "Google," where he served until 2008 as the chief information officer. Since leaving Google, Douglas has worked as president of Digital Music and COO of New Music for EMI Recorded Music, and he's now heading his own start-up company. And he's only just turned forty.

When we were preparing to talk to Douglas, we noticed that in his previous interviews he'd often used analogies to make his points. We asked him whether detecting similarities was a key element of his thinking. "Absolutely, I love analogies. They're my bread and butter . . . if you'll let me use an out-of-date analogy. Often, the things I'm interested in doing arise from some analogy I come up with. I don't understand what I'm doing until I have a few analogies to describe it. The first element of my storytelling is asking, 'What's the story going to be like?' Then, for each major point, I brainstorm analogies."

When we asked him if he preferred to solve problems using interdisciplinary approaches and multiple perspectives, Douglas responded, "Yeah, totally. I just think there's lots of different views on problems, and that by seeing more than one you're better off."[3]

Clearly, Douglas is equipped with remarkable I-strengths. As you'll see after you've read parts 5 and 6, he also has remarkable N- and D-strengths. So what about M-strengths, or the kind of spatial reasoning ability that many people think of as the characteristic dyslexic strength? Douglas laughed when we asked him. "I'm abysmal at spatial reasoning. If I close my eyes I can't tell you which way my office door is from where I'm sitting. But what I can do is tell you that it's nine turns from my office to my house, and they're on average three blocks apart. So I don't know where my house is from here, but I guarantee I can get you there."

Turns out that individuals with dyslexia with powerful I-strengths have many ways of making connections.

CHAPTER 14

Key Points about I-Strengths

In the last few chapters we've discussed the critical role that Interconnected reasoning plays in the thinking of many individuals with dyslexia. Key points to remember about I-strengths include:

- The ability to spot important connections between various kinds of information is an important—and possibly even the most important—dyslexic advantage.
- I-strengths include the abilities to see relationships of likeness and "togetherness"; connections between perspectives and fields of knowledge; and big-picture or global connections that create heightened abilities in detecting gist, context, and relevance.
- I-strengths appear to be enhanced in individuals with dyslexia because their brain microcircuitry is biased toward the creation of highly interconnected, long-distance circuits that favor top-down, global processing and the recognition of unusual relationships.
- This structural and cognitive bias creates a trade-off between enhanced I-strengths and challenges with fast, efficient, and accurate fine-detail processing.
- Dyslexic learners with prominent I-strengths can be greatly aided in

learning by performing a few simple steps, including providing summaries or overviews of longer reading passages, pre-learning key vocabulary, providing information about the practical importance and usefulness of material being taught, tying in new information with things pre-existing knowledge, and beginning courses or units with an overview of the goals, "the big picture," and outlining the lesson plan that will be followed.

Interconnected Reasoning in Real Life

Let's close by examining how I-strengths have helped another highly talented individual with dyslexia: philosopher, thought leader, author, and corporate CEO Dov Seidman. As a younger student, Seidman struggled greatly in school. In fact, he jokes that his only two A's in high school were in physical education and auto shop. Then, in college, his approach to learning was revolutionized by his serendipitous encounter with philosophy. As Seidman told us, "I fell in love with philosophy. With my professors' encouragement, philosophy helped me overcome my dyslexic challenges. Unable to read hundreds of pages, philosophy rewarded me for the careful consideration of one idea, and my disability transformed into a strength."

Philosophy, at its roots, is the search for gist—for context and connections of all kinds. It focuses primarily on big-picture views rather than fine details, and Seidman found both that his mind was perfectly suited to philosophy and that philosophy was developing his mind in exciting ways.[1] The methods and disciplines he learned through its study taught Seidman to see "through the words to the ideas that lay beneath them" and to focus on general principles—what we've been calling gist.

Seidman found he had a natural ability to create frameworks for understanding the world, addressing how we as humans seek alignment in our relationships and among competing interests. Seidman went on to earn simultaneous bachelor's and master's degrees, summa cum laude, in philosophy from UCLA. He later earned a B.A. with honors in philosophy, politics, and economics from Oxford University.

Seidman then decided to test his blend of skills and interests at Harvard Law School, and immediately following he took a job in a large law firm. In his book, *How: Why HOW We Do Anything Means Everything . . . in Business (and in Life)*, Seidman describes what happened next: "Toiling away in the law library, it dawned on me that someone somewhere had researched the very issue I was working on, and inevitably knew more about it than I did (which was zero). I saw an opportunity to make legal knowledge accessible to a large number of people in business at a low price."[2]

Let's pause for a moment to review what occurred. Seidman was performing the kind of routine legal research that thousands of young attorneys perform every day without ever deeply questioning what they're really doing or why they're doing it. But Seidman was different. His mind had become attuned to looking more deeply at the work he was engaged in, and this "philosophical habit" led him to step back from his task to search for its gist and context.

Seidman's search revealed two things: how redundant it was for him to research a topic that countless lawyers had previously researched, and the opportunity this redundancy provided to create a company that offered expert legal research on a wide range of topics. That's how Seidman's company, LRN, was born.

As LRN grew, Seidman made another important discovery: "[T]he core of our efforts lay in helping our clients put out fires by responding to legal challenges that had already arisen. I began to believe that we could be of better service by helping them . . . prevent these legal problems from arising in the first place." In other words, rather than simply focusing on their immediate need for research, Seidman again stepped back to search for the underlying sources of that need.

As Seidman analyzed the causes of his clients' problems, it struck him that they weren't ultimately best understood from a legal, regulatory, or compliance perspective but from the perspective of organizational and individual behaviors and the values that root and guide those behaviors. As Seidman told us, "the ancient philosopher Heraclitus once said character is fate, and corporations are analogous to individuals in having a character, patterns of behavior, and the capacity to earn a reputation." These realizations led Seidman to conclude that the key to preventing corporate legal problems was to teach corporations to do everything they did—from managing employees to producing products to dealing with clients—the same way a person of good character would. In other words, Seidman combined the perspectives of philosophy, law, and business to create a new way of viewing the interactions, behaviors, interests, and obligations of individuals and corporations, which would lead to better corporate behaviors and to better business outcomes.

Seidman's impressive I-strengths—his abilities to detect relationships between concepts and events, to see the benefits of adopting different perspectives, and to look behind what's immediately apparent to see the deeper and broader significance—have all been crucial to his success. Seidman has been named one of the "Top 60 Global Thinkers of the Last Decade" by the *Economic Times* and "the hottest advisor on the corporate virtue circuit" by *Fortune* magazine, and his company has helped more than 10 million people in more than five hundred companies worldwide. That's the power of Interconnected reasoning.

PART V

N-Strengths

Narrative Reasoning

CHAPTER 15

The "N" Strengths in MIND

Anne was "a consistently poor reader" until well into adulthood. Like many struggling readers, her memories of school are highly negative: "School was torture. School was like being in jail. It was captivity and torment and failure."[1] Though she dearly loved stories and spent hours flipping through picture books, her poor reading skills kept her from drawing more than a bare sketch of the "action and incident" described on the page. Instead, it was through books read aloud at school and home, and the radio dramas and movies she enjoyed, that she developed a love for the rhythm and flow of language.

Anne struggled with reading throughout elementary school, but writing grew easier. From fifth grade on, she wrote adventure stories and plays for her classmates. They responded enthusiastically and overlooked her spelling errors. Unfortunately, Anne found no way to turn her writing talent into classroom success.

It wasn't until her freshman year of high school that she finally read well enough to appreciate the actual words in the books she read. "The first novel that I recall truly enjoying and loving for its language as well as its incident was *Great Expectations* by Charles Dickens. . . . The other novel . . . was Charlotte Brontë's *Jane Eyre*. . . . I think it took me a year to consume these two books. It might have taken two years. . . . [I]t was a slow go."

Despite these challenges, Anne's love of literature and writing continued to grow. When she went off to college, she decided to major in English. Unfortunately, she soon had to abandon this plan because she was still so "severely disabled as a reader" that she couldn't complete the assignments for her classes. Getting through even one of Shakespeare's plays in a week was virtually impossible for her, and the written work was equally difficult: "[I] barely got by . . . because I wasn't considered an effective writer. The one story I submitted to the college literary magazine was rejected. I was told it wasn't a story." Anne's spelling, too, remained a problem. As she told us, "I can't spell to this day. I don't see the letters of words, I see the shapes and hear them. So I still can't spell. I'm always looking up spelling and making mistakes."

Anne began looking for another subject where she might find more success. She was passionately interested in the great ideas and beliefs that shaped the modern world and wanted to form a "coherent theory of history." She considered majoring in philosophy, but here, too, she was hindered by her poor reading. Anne found that she "could only make it through the short stories of Jean-Paul Sartre and some of the works of Albert Camus. Of the great German philosophers who loomed so large in discussion in those days [during the early 1960s], I could not read one page." Instead, Anne opted for a degree in political science, where she was able to grasp the key concepts almost entirely from lectures. She earned her degree in five years.

After graduation, Anne remained drawn to writing and literature. At age twenty-seven she returned to school to study for a master's degree in English, which she earned in four years. "Even then I read so slowly and poorly that I took my master's orals on three authors, Shakespeare, Virginia Woolf, and Ernest Hemingway, without having read all of their works. I couldn't possibly read all of their works."

Fortunately, Anne could still write, and shortly after earning her master's degree she began work on a new novel. One of the primary themes of that novel was the experience of being "shut out" from life and the fulfillment of dreams—an experience Anne knew well from being "shut out of book learning." Three years later that novel was published, and it became a phenomenal bestseller. Anne followed that first novel, which she entitled *Interview with the Vampire,* with twenty-seven more. Together they've sold over 100 million copies, making Anne Rice one of the bestselling novelists of all time.

Narrative Reasoning: The Structure of Experience

You might think it's extremely unusual for such a talented and successful writer to have trouble with reading and spelling. You would be wrong.

Many highly successful writers have faced dyslexic challenges with reading, writing, and spelling, yet have learned to produce clear and effective prose. Even limiting our selection to contemporary writers whose dyslexic symptoms can be clearly confirmed, the list of successful dyslexic authors is impressive and includes such notables as:

- Pulitzer Prize–winning novelist (*Independence Day*) Richard Ford
- Bestselling novelist (*The World According to Garp, A Prayer for Owen Meany*) and Academy Award–winning screenwriter (*The Cider House Rules*) John Irving
- Two-time Academy Award–winning screenwriter (*Kramer vs. Kramer, Places in the Heart*) Robert Benton
- Bestselling thriller writer Vince Flynn, whose novels have sold over 15 million copies in the last decade
- Bestselling mystery writer, screenwriter (*Prime Suspect*), and Edgar Award winner Lynda La Plante
- Bestselling novelist Sherrilyn Kenyon (who also writes under the name Kinley MacGregor), whose novels have sold over 30 million copies[2]

We're not mentioning these outstanding creative writers just to encourage and inspire you with their remarkable achievements. Nor are we merely suggesting that dyslexic processing can be helpful for creative writing, though for reasons we'll discuss shortly we also believe this to be true. Instead, we're focusing on these talented writers because we believe they reveal something important about dyslexic processing *in general*—not just for dyslexic writers, but even for many individuals with dyslexia who never write at all. What these authors illustrate is the profoundly narrative character of reasoning and memory that many individuals with dyslexia possess. This Narrative reasoning is the N-strength in MIND.

N-strengths are the ability to construct a connected series of "mental scenes" from fragments of past personal experience (that is, from episodic or personal memory) that can be used to recall the past, explain the present, simulate potential future or imaginary scenarios, and grasp and test important concepts.

While many individuals with dyslexia might not instinctively regard their thinking as "narrative" in style, we'll show you the ways in which the memory and reasoning styles that many individuals with dyslexia display are, in fact, profoundly narrative. We'll also show you the many amazing ways that N-strengths can be employed.

CHAPTER 16

The Advantages of N-Strengths

N-strengths draw their power from a kind of memory known as episodic or personal memory. To understand how episodic memory supports the N-strengths, it will be helpful to briefly review how the memory system as a whole is structured.

The memory system can be divided into two main branches: *short-term* and *long-term memory* (see figure 1). Short-term memory—which contains

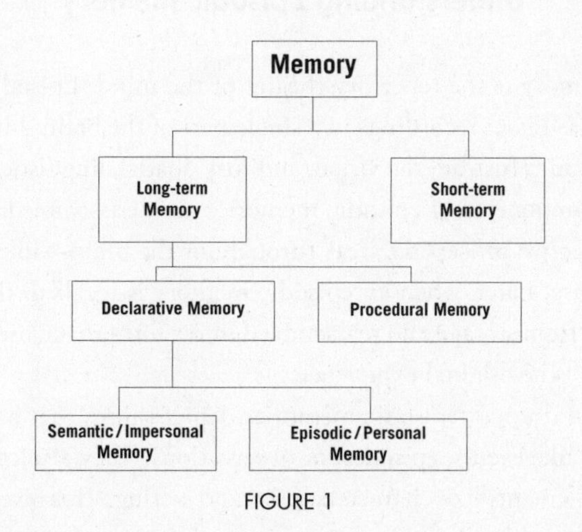

FIGURE 1

both *short-term* and *working memory*—is responsible for "keeping in mind" the information you're using right now. Long-term memory stores information you can retrieve and use later.

Long-term memory, which will be our focus in this chapter, also has two branches: *procedural memory* and *declarative memory.* Procedural memory holds the "procedures and rules" that help us remember how to do things. Declarative memory stores "facts about the world."

Declarative memory can be further divided into *episodic* and *semantic memory.* Episodic memory (also called *personal memory*) contains factual memories in a form that simulates events, episodes, or experiences. Semantic memory stores facts as abstract and impersonal data, stripped of context or experience.

Many facts about the world can be recalled either as episodic or semantic memories. For example, the fact that "tears taste salty" can be recalled as an episode you've experienced or as a fact you simply know without remembering anything about the episode in which you learned it. We'll focus on episodic memory because it underlies N-strengths and it is the preferred way of storing factual knowledge for many individuals with dyslexia.

Understanding Episodic Memory

Episodic memory is the repertory theater of the mind. Episodic memories aren't stored as intact recordings in a single part of the brain—like old movies in a film vault. Instead, the visual, auditory, spatial, linguistic, tactile, and emotional components of episodic memories are disassembled, then stored in their respective processing areas throughout the brain—like stage props in a warehouse. Later, when an episodic memory is recalled, these "props" are retrieved from storage and reassembled into a form that closely resembles (or "restages") the original experience.

Like most dramas, episodic memories depict things that happen or are experienced, like events, episodes, or observations. They also contain traditional story elements like characters, plot, and setting. This gives them their narrative or storylike character.

This process of restaging mental "scenes" from fragments of past personal experience is an extremely powerful way of recalling facts about the world. We can get a glimpse of this power by inspecting the recollections of an individual with an extremely rich episodic memory, novelist Anne Rice. One of the most remarkable things about Anne's autobiography, *Called Out of Darkness*, is the vividness and clarity of her memories from childhood, as shown in this description of a walk she often took with her mother:

> We left our house . . . and walked up the avenue, under the oaks, and almost always to the slow roar of the passing streetcars, and rumble of traffic, then crossed over into the Garden District. . . . This was an immediate plunge into a form of quiet. . . . I remember the pavements as clearly as I remember the cicadas singing in the trees; some were herringbone brick, very dark, uneven, and often trimmed in velvet green moss. . . . Even the rare stretches of raw cement were interesting because the cement had broken and buckled in so many places over the roots of the giant magnolias and the oaks.

This description is so clear—so rich in atmosphere and sensory detail—that it draws the reader right to Anne's side as she makes that journey. Yet when she wrote it, Anne was describing walks she'd taken nearly sixty years earlier.

As powerful as this "restaging" function is, recalling the past is only one of episodic memory's many functions. As the White Queen in Lewis Carroll's *Through the Looking-Glass* quite rightly observed, "It's a poor sort of memory that only works backwards." Episodic memory escapes this criticism because it not only helps us recall the past, but it also helps us understand the present, predict and envision the future, mentally simulate planned actions or inventions, imagine events we haven't witnessed or that are fictitious, solve problems, navigate, and create narratives that can persuade or enlighten others.

To help explain the many functions of episodic memory, we spoke with Dr. Demis Hassabis, a neuroscientist who's played a key role in this still new and rapidly advancing field.[1] In 2007, Dr. Hassabis coauthored a groundbreaking paper with his colleague Dr. Eleanor Maguire that described the remarkable versatility of the episodic memory system.[2] Their paper, which was voted one of the ten most important scientific papers of the year by the prestigious journal *Science*, introduced the term *scene construction* to describe the core process by which episodic memory works to perform its many functions.

When we spoke with Dr. Hassabis from his office in London, he explained this process of *episodic construction* in the following way. "Episodic memory reconstructs things you've previously experienced from the remembered elements you've acquired through your experiences in life. For example, say you walk through a beautiful garden or park, and you see a beautiful rose, and you smell the rose: all those elements of experience become components in your memory. Later, when you want to recall what you've experienced, you reassemble those components in a way that looks familiar. You may get some of the details wrong because memory is often inaccurate, but to the extent that you're right, that's an accurate reconstruction of an episodic memory."

We then asked Dr. Hassabis to explain some of the additional functions of episodic memory. He responded, "Recently we've found that using scene construction to recall the past is just one small part of a much bigger system, which we call the *episodic simulation* system. Episodic simulation is very powerful because it allows memory to be used *creatively*. With creativity you assemble the same kinds of memory elements that you use to recall the past, but rather than reconstructing something you've experienced before, you combine the elements in *new ways* to construct a whole that's entirely novel because it contains unprecedented connections between the elements. In other words, creativity uses the same construction process that you use to reconstruct memories, but the construction is creative because it results in something you've never experienced before. The process is similar, but the outcome is entirely new."

This creative recombination system can be used as a kind of mental laboratory to simulate what might happen given certain starting conditions or circumstances. According to Dr. Hassabis, "This episodic simulation function is very valuable in a lot of fields, including things like financial forecasting, or designing computer games and imagining how players will play them, or thinking about a film scene and how it might play out."

Given the many uses of this episodic simulation system, the bias for *episodic* rather than *semantic* memory, which many individuals with dyslexia show, has implications that go beyond memory to reach the very heart of the reasoning process. Dyslexic individuals with prominent N-strengths often reason by mentally simulating potential events or actions, then "observing" how these simulations "play out," rather than reasoning abstractly using definitions or formulas stripped of context. These simulations are based on information they've gathered from real experience rather than on abstract principles.

Scene-Based versus Abstract, Noncontextual Knowledge

We've found that a large majority of individuals with dyslexia show this preference for episodic over semantic memory for most tasks, and it shows up in various ways in both clinical settings and real life. One way that it presents is as a tendency to store conceptual and verbal knowledge as *scene-based depictions or examples* rather than *abstract verbal definitions.*

Often as a part of our testing sessions we'll ask individuals to define terms or concepts. While most individuals respond with abstract dictionary-type definitions, individuals with dyslexia often respond with examples, illustrations, anecdotes, or descriptions of uses or physical features. For example, when we ask individuals with dyslexia to define the word *bicycle*, they're more likely than nondyslexics to respond with an analogy (e.g., "It's like a motorcycle, but you make it go yourself") or a description (e.g., "It's a thing with a seat, two wheels, handlebars, and pedals that you make go by pushing the pedals with your feet"), as opposed to an abstract definition (e.g., "It's a human-powered, two-wheeled, transportation device"). The same is true when we ask individuals with dyslexia to define a concept that's inherently abstract, like "fairness." Individuals with dyslexia are more likely to respond with an example ("It's like when you're playing a game and you wouldn't want to make someone else do something you wouldn't want to do yourself") than an abstract definition (e.g., "It means everyone should be treated the same" or "It means you get what you deserve"). This reliance on scene-based *depictions* of facts rather than abstract or noncontextual *definitions* reflects a greater reliance on episodic rather than semantic memory, and many of the older dyslexic individuals with whom we've spoken confirm that this pattern is characteristic of their thought.

When we ask these older individuals with dyslexia to tell us more about their thinking style, they often also describe another feature that relates to episodic memory. When thinking of a fact or concept, they typically find that the concept is not represented in their mind by a single generalized depiction of that concept, but rather by a series of distinct examples through which they can mentally "scroll." The concept is understood as the complete collection of all these examples, and while it centers around the most common or representative examples, it also includes the outliers. Jack Laws gave us an especially good description of this conceptual style. When we spoke with him, he mentioned that he'd found he could much more easily distinguish between different animals of the same species than most people—for example, between different crows or robins. So we asked him what popped into his head when we mentioned the concept "robin": just one idealized robin or a whole series of different robins? He answered without hesitation, "Different robins, definitely. My mind starts jumping to robins that I've experienced, rather than a single generalized robin, or the Platonic ideal of the robin." When he drew a robin for his field guide, he drew the single robin that best represented the features of the whole group—but it was a *particular* robin rather than an idealized generalization. In other words, he drew from an episodic rather than a semantic memory.

When we mentioned these dyslexic memory preferences to Dr. Hassabis, he responded, "That's very interesting because it relates to an important trade-off between episodic memory and semantic memory that people in the memory field have been thinking about. Both of these types of memory are critically dependent upon the hippocampus [which, as we mentioned before, plays critical roles in memory formation and recall]. But the interesting thing is—and this is really not written anywhere, but it's the sort of thing people at the cutting edge are thinking about at the moment—if you want to be very good at episodic memory, you want your hippocampus to engage in a process called *pattern separation.* What pattern separation involves is this: Suppose you experience something new, and even though it's quite similar to something else you've experienced, you want to remember it as a distinct event—for example, what I did for lunch yesterday as opposed to three days ago, even though maybe I had lunch with the same people, and in the same place, and lots of the elements were in common. One of the functions the hippocampus performs is to keep those similar memories separate. It actually makes them more divergent than you would expect—and that's exactly what you'd want if you wish to have a good episodic memory.

"In contrast, if you want to be good at learning semantic facts that generalize and are true across multiple experiences—for example, the fact that Paris is the capital of France—then you don't really care about the specific episode you learned that in. What you care about is basically just the 'fact nugget.' The surrounding context in which you learned that fact simply isn't relevant to that piece of information. In that case you want something else the hippocampus does, which is called *pattern completion.* Pattern completion is a process that unites divergent things. So, let's say you heard a particular fact in several different lessons. What's actually important is *that fact* and not the different contexts in which you encountered that fact. So pattern completion solidifies your memory of the fact that was heard on each occasion but eliminates any record of the differences in the way you experienced it.

"Now, if the hippocampus is responsible for doing both of those things, then perhaps what you're seeing with these dyslexics is that some of the same brain wiring differences that cause them to be dyslexic also predispose them to favor pattern separation over pattern completion. That would make them very good at remembering things that have happened to them and at episodic memory. This greater diversity of separated patterns might also make them better at spotting unusual connections between facts that people who are not dyslexic wouldn't make."

This last observation is very important because it relates to our discussion of I-strengths in part 4. There we suggested that creativity may be enhanced in individuals with dyslexia because they are predisposed to making broader neural circuits that both create a greater breadth, diversity, and novelty of connections and enhance the perception of gist and context. Pattern separation also empowers the dyslexic mind because it "stocks" the mind with a greater number of separated patterns that may be used to make novel connections. In short, individuals with dyslexia may have a double helping of cognitive features that enhance their ability to make diverse and more creative connections.

These features are ideal for producing minds with powerful narrative abilities. What could be more helpful for a storyteller than a mind stocked with an endless array of different characters and experiences and scenarios; disposed to spot new connections, associations, patterns, and nuances between them; and wired with the ability to unite it all into a single great narrative by seeking a higher-order context or gist?

Thinking in Stories: A Common Dyslexic Strength

This highly creative narrative thinking style often displays itself in a tendency to think and convey information in story form. We first noticed this tendency during our testing sessions when we asked individuals to describe a picture called "Cookie Thief."[3] It shows a woman standing in the foreground, at the front plane of the picture, drying a dish with a towel and looking out rather vacantly toward the viewer. Behind and to her left, water is overflowing from a sink where the tap has been left on, and it's beginning to collect in a puddle on the floor. To her right—and clearly unbeknownst to her—a young boy is standing on a stool, reaching high into a cabinet for a cookie jar that sits on the top shelf. Next to the stool, a girl is reaching up eagerly to receive a cookie. Neither seems aware that the stool is tipping and the boy is about to fall.

Most viewers find the events in this scene rather trivial and implausible—especially the actions of the woman, who seems bizarrely detached from the chaos around her—so they make little effort to reconcile the various events into a single coherent story. Instead, they simply describe the most obvious features of the picture. Over time, however, we found that a small number of viewers would propose additional details in an effort to reconcile the picture's seemingly irreconcilable elements. Most often the additional detail would involve a person or an object "in front" of the plane of the picture (like the father or a TV) that was distracting the woman's attention and causing her to ignore the sink and the children. Remarkably, nearly all the individuals who proposed these extra elements were dyslexic, and the solutions they proposed were clearly aimed at identifying the gist to provide a coherent explanation for the action in the picture.

We also found that the individuals with dyslexia were more likely to use conventional storytelling techniques to describe the picture. At younger ages these include formulaic openings like, "One day while she was washing the dishes . . . ," or "Once upon a time . . . ," but even the older individuals with dyslexia were more likely to give the characters names, lines of dialogue, distinctive personality traits, senses of humor, motivations, and personal and family histories. We've found that many individuals with dyslexia use these kinds of narrative, personal, or episodic elements in all sorts of descriptive tasks, and that their descriptions often contain elements like analogies, metaphors, personalizations or anthropomorphizations, and vivid sensory imagery.

This bias for Narrative reasoning can also be seen in the professional lives of many individuals with dyslexia, who use their N-strengths in all sorts of ways. The following are several examples of dyslexic individuals in fields other than creative writing who have flourished using narrative skills.

Duane Smith is professor of speech and director of the public speaking team at Los Angeles Valley College—a school he once failed out of as a student due to his dyslexia-related challenges. As he told us, "My whole life has been about stories, and telling stories, and I stress the importance in my public speaking class about telling stories. Half of what we do in forensics competitions is to perform stories, but for me literally everything is about stories." Prior to becoming a professor, Duane had a successful career in sales, where he also found his narrative skills to be invaluable.

When we described "episodic memory" to Duane, he laughed in recognition. "If I hear a song, or smell something, or see an article of clothing or a car from a particular year, I can immediately imagine a scene on a particular day, or event. It drives my wife crazy because we'll be listening to the radio, and I'll talk about how it takes me back to 1985 when I was standing with a group of buddies at In-N-Out Burger on a Saturday night listening to that song, and what we were talking about, and she'll say, 'Can't you ever just listen to the song?'" By contrast, Duane told us that he remembers almost nothing in abstract, noncontextual form: "The only things I remember are experiences and examples and illustrations."

Law professor David Schoenbrod recalls that when he was a junior in high school, his English teacher told his parents, "David is literate in no language." That's a problem he's long since overcome, as readers of his four highly regarded books on environmental law will attest. To what does David attribute his success as a litigator? As he told us, "It seems to me that my strength as a lawyer was being able to tell a story. I had a colleague early in my career who told me that the way you win a case is by telling a story in a way that makes the judge want to decide your way. And I've always felt that I was good at that. . . . I like storytelling, and it came readily to me."

Entrepreneur and cognitive psychologist Douglas Merrill attributes his survival in school and his mastery of math techniques primarily to his use of narrative strategies. "I always think in stories. . . . I spent most of my time [as an adolescent] reading or telling stories, or playing fantasy games around stories like Dungeons and Dragons.

"I ended up at Charles Schwab, and Charles—who goes by Chuck—is dyslexic; and he sits in meetings with his eyes closed, listening to people talk, and he never reads the handouts in advance, and it's pretty clear that all he's doing is listening and thinking. And he tells these great stories about what customers are going to want. I found that incredibly freeing because for the first time I thought, 'This stuff I do well at is valuable, as opposed to the stuff I do badly at, which seems to be what everyone else thinks matters.'"

After leaving Google, Douglas worked briefly as president of New Music at EMI Recorded Music. "I thought one of the problems that industry had

was that it didn't know anything about itself. So I spent a lot of time trying to figure out what's actually happening in the music industry. The most direct way to do that is with math, and I suck at math, so instead I would read economic articles and surveys, and I would make notes on yellow stickies, then stick them on the wall. Then once a week I'd skim those stickies and move them around, and what I ended up with was a story."

In short, N-strengths can be useful in any job or task where past personal experiences can be used to solve problems, explain, persuade, negotiate, counsel, or in some way form or shape the perspectives of oneself or others.

CHAPTER 17

Trade-offs with N-Strengths

In addition to the many abilities, N-strengths can also bring trade-offs. The most important—and most common—trade-off is reflected in the following comment: "Sammy never remembers anything from school. He forgets what's been taught, and whether there's an assignment or a test. And when we ask him to do something at home, it's always in one ear and out the other. His memory is horrible! Only, the strange thing is, Sammy's also our family historian. He can remember what we've done on every vacation, and who gave what present at his brother's birthday party five years ago, and what kind of pet every kid in his class has. So why can he remember all those kinds of things but he can't remember his times tables or the names of the state capitals?"

This classic, but seemingly paradoxical, description of the "family historian with the poor memory" is one we hear from countless families. It becomes easier to understand when you remember the different types of memory we described at the start of the previous chapter.

All individuals, whether dyslexic or not, show a distinctive blend of strengths and weaknesses in episodic, semantic, and procedural memory, and this blend greatly affects their learning and memory styles. Like Sammy, many individuals with dyslexia have a much stronger episodic than semantic memory and a relative weakness in procedural memory (as discussed in chapter 3). Dyslexic individuals with this memory style are typically very good at remembering things they've done or experienced, and often also at remembering stories that they've heard or information that's been embedded in a narrative context. However, they're much weaker at remembering "bare semantic facts"—or facts that are abstract, impersonal, and devoid of context.

Anne Rice is a perfect example of an individual with this memory pattern. As we mentioned earlier, Anne has a phenomenally good memory for episodic and personal details—that is, for things she's experienced; yet Anne also has a poor memory for abstract, impersonal facts. As she told us, "I don't think abstractly at all. Everything is image and narrative with me. I can't remember numbers at all and make huge errors, sometimes doubling prices or amounts as my memory of them gets hazy."

It's critical to identify students with dyslexia who show a primarily narrative processing and episodic memory style, because their N-strengths can provide the key to unlocking their learning potential. This is true both for the ways they take in and the ways they express information.

When taking in new information, students with dyslexia who show a strong episodic memory bias and narrative processing style will typically learn much better if general or abstract definitions are supported by scene-based examples or depictions. When information is embedded in a context that the student finds meaningful and familiar, and which incorporates experiences, cases, examples, stories, or personal experiences (including humor, participation, novelty, "strangeness"), many students with dyslexia will learn it more quickly and retain it more durably.

These points are reinforced by the following experience, shared by the mother of a dyslexic child. She told us that as a student she'd always excelled at remembering facts, definitions, and formulas, while her dyslexic husband had always struggled in these areas. She'd been the honors student while her husband had barely made it through school. So she naturally assumed she'd be a better tutor for their dyslexic son. She was surprised, then, to discover that her husband was a *much* more effective teacher, especially for concept-rich subjects like history, social studies, and science. Eventually she realized this was because her husband taught almost entirely using examples, cases, and analogies, while she tried to "trim the fat" from her lessons and present only the bare minimum of "simple facts" so their son would have less to memorize. However, it wasn't the *quantity* of information that their son struggled with but the *form*. He could hold on to facts that were embedded in a meaningful story or context but quickly forgot facts that lacked context or significance. This is a common experience for students with dyslexia.

Dyslexic students with Narrative reasoning styles also face important challenges when trying to express their ideas. Since their conceptual knowledge is often stored in cases, images, or narratives rather than in abstract principles or definitions, when asked to answer questions on exams or assignments or even orally in class by stating the relevant abstract or general principles, they may respond instead with stories or examples. As a result, their answers may appear loose and unstructured. They may seem to "talk around" their answer and appear to have difficulty "getting to the point." Douglas Merrill shared one example of how this happened to him.

"When I took my qualifying exams in graduate school, I was asked a question about the development of cognition; and I was supposed to start with Piaget, then go to Erikson, then go to modern cognitive problem solving. I understood all the important concepts, but I hadn't really been able to memorize all the little details, so instead I wrote a story about the different developmental paths of two people. I covered all the right concepts, but they failed me on that question because I didn't give them all the specific details that they wanted."

As we mentioned earlier, this dyslexic tendency to think in examples or stories rather than in abstract definitions can also result in the loss of points on standardized tests, including (and perhaps especially) the vocabulary portion of IQ tests.

Schools and exams often treat abstract facts and principles as if they were the only forms of knowledge that really "count"; they assume that if students can't memorize and regurgitate facts in their "purest" and most noncontextual form, then they don't really *know* them. While abstract definitions are important and useful, we must not undervalue knowledge that is embedded in experiences, stories, cases, or examples. Such *case-based knowledge* is highly valuable in its own right, and it is more easily mastered than abstract information by many students with dyslexia.

It's also important to recognize that individuals with a largely narrative or case-based reasoning style will often show a very different trajectory of cognitive development from individuals with a more abstract or semantic reasoning style. This is particularly true of the growth of their conceptual knowledge. At younger ages, individuals who store concepts as cases and examples can appear concrete because they have few cases and experiences to reason with, so they may seem "stuck on" overly specific cases when asked to think about a broad concept. Early on, such children often have more difficulty generalizing their knowledge than their peers. Fortunately, as their experience increases, so will the fluidity of their thinking. In fact, once they have accumulated a broader set of experiences, they will often be *less* concrete than others because their concepts include a wide range of cases rather than a single generalized principle. This also makes them less likely to mistake an abstraction or generalization for a full description of reality.

Finally, because narratives speak so powerfully to individuals with dyslexia who possess prominent N-strengths, it's crucial that *the narratives we tell them about dyslexia* are both accurate and appropriately hopeful. One of our chief goals in writing this book is to correct the common and deeply misleading narrative that dyslexic differences are primarily, or even entirely, dysfunctions. The "story" we should read in the lives of the individuals with dyslexia isn't a tragedy; it's an exciting story filled with hope, opportunity, and promise for the future.

CHAPTER 18

N-Strengths in Action

Let's look in detail at the many uses to which one highly talented individual with dyslexia has put his N-strengths.

Blake Charlton was diagnosed with dyslexia midway through first grade. Despite being passionately fond of hearing and telling stories, Blake made little progress with reading and writing, and he struggled with basic math.

Blake spent two years in a special ed class, where he began to make progress. He enjoyed feeling like the "smart kid" in class and was pleased when he could finally start putting stories down on paper.

By fourth grade Blake had progressed enough to be mainstreamed back into the regular education class. His sense of accomplishment quickly vanished as he went from being the "smart kid" to the "class failure." Only his skills in sports and drama—which earned the admiration of his classmates—allowed him to keep a positive self-image and respond to his classroom setbacks with a determination to improve.

In middle school this resolve helped Blake boost both his reading speed—largely by consuming the fantasy novels that gripped his imagination—and his pace of work. However, Blake continued to make "silly" mistakes in writing and math, and they chipped away at his grades.

Blake finally experienced a breakthrough when he was allowed to use a calculator and spell-checker for his work. His grades shot up, and so did his self-esteem. "Suddenly, I was a geek again!" Blake enjoyed being recognized for his intelligence, and he began to apply himself with even greater determination. With accommodations in place for his College Board exams, Blake did well. So well, in fact, that he was admitted to Yale University.

Though Blake retains fond memories of Yale, he also remembers college as a time of terror. He was so concerned that his remaining dyslexic difficulties would defeat him that he compensated by spending "every waking moment of the day" on his studies. Fortunately, Blake received invaluable assistance from Yale's Resource Office on Disabilities. The staff helped him obtain classroom accommodations such as a keyboard for in-class work, extra time on tests, and out-of-class assistance with proofing papers, planning, and scheduling.

Because he'd long dreamed of becoming a doctor, Blake began to take science courses, and he discovered that he had a special knack for chemistry. Blake particularly excelled in organic chemistry, with its heavy emphasis on three-dimensional spatial reasoning, and he actually earned the top grade in this very difficult and competitive class. Blake also did well in inorganic chemistry, because even though his rote memory was weak, he was still able to remember an astonishing number of facts about the chemical elements in the periodic table by creating fanciful stories about them. He gave the elements personalities, past histories, motivations, and goals, and these narrative details helped him remember their "behaviors" and their positions in the rows and columns of the table. Blake used similar narrative-based memory strategies in his other courses, too. As a result of these successes, Blake told us that for the first time in his life he felt truly "intellectually talented."

Blake might have been happy as a chemistry major had his love of stories not been so strong. Instead, he majored in English, and with persistent effort his writing skills began to blossom. In fact, he won two writing awards while at Yale.

Following his graduation in 2002, Blake took a job as an English teacher, learning disabilities counselor, and football coach, then returned home to care for his father, who was battling cancer. In the few spare moments he somehow managed to find, Blake wrote stories about the imaginary worlds he'd dreamed of all his life, while continuing to dream about becoming a doctor.

In 2007 Blake finally entered Stanford Medical School. At the same time he also signed a three-book deal with a publisher specializing in fantasy fiction. Blake's first novel, *Spellwright*, was published in 2010. Fittingly, *Spellwright* is the story of a magician-in-training with dyslexia who must solve the riddle of his own "cacography"—or inability to handle text-based spells without "corrupting" them—to prevent the triumph of evil over good. It's an absolutely thrilling read, and the elaborate system of magic Blake creates is astonishing in its inventiveness.

When we spoke with Blake, he was taking time off between his second and third years of medical school to teach creative writing to first-year medical students (as a way of encouraging their use of Narrative reasoning in clinical medicine) and to publish an analysis of literary narratives related to medicine (such as Tolstoy's story *The Death of Ivan Ilyich*). During our conversation, Blake told us how useful he'd found narrative-based memory strategies for dealing with the overwhelming amounts of memorization he'd faced in his first two years of medical school. He even shared several of the stories he'd developed to help remember the branches of arteries and nerves. Narrative reasoning is clearly still a dominant theme in his life.

Oh, and in his "spare time" Blake was also completing the second novel in his planned trilogy, *Spellbound*. Like his first, it combines elements of Blake's experiences with dyslexia and the remarkable system of magic he created. Of course, we shouldn't be surprised to find that someone who can transform himself from a special education student to a Phi Beta Kappa graduate of Yale knows a thing or two about magic.

CHAPTER 19

Key Points about N-Strengths

Narrative reasoning plays a key role in the thinking of many individuals with dyslexia. Key points to remember about N-strengths include:

- Many individuals with dyslexia show a profound difference between their powerful episodic (or personal) memories for events and experiences and their much weaker semantic (abstract or impersonal facts) and procedural memories.
- Episodic memory has a highly narrative or "scene-based" format in which concepts and ideas are conceived or recalled as experiences, examples, or enactments rather than as abstract, noncontextual definitions.
- The episodic construction system can use fragments of stored experience not only to reconstruct and remember the past but also to imagine the future, solve problems, test the fitness of proposed inventions or plans, or create imaginary scenarios and stories.
- Episodic construction and creativity can be closely linked.
- Individuals who rely on episodic or narrative concepts rather than abstract, noncontextual facts will typically reason, remember, and

learn better using examples and illustrations rather than abstract concepts or definitions.

- Many individuals with dyslexia will learn and remember better by transforming abstract information into narrative or case-based information through the use of memory strategies or stories.
- Many individuals with dyslexia enjoy (and are skilled in) creative writing even though they may have difficulty with formal academic writing or reading; so teachers should look carefully for signs of narrative ability in students with dyslexia, and they should help talented individuals with dyslexia further their abilities through the use of appropriate tutoring and accommodations.
- Narrative approaches can be useful for all sorts of occupational and educational tasks, not just creative writing.

Let's close by looking at a remarkable young writer with dyslexia who's still at the beginning of her career. This eleven-year-old girl came to see us from England. When we asked if we could share her work, she agreed but asked to be called by the nom de plume Penny Swiftan. The following is taken from a story she wrote shortly before visiting us:

> For a moment the stars blazed bright, and Lady stared in amazement and wonder at the sight before her. The glade was ringed by foxgloves, oak trees and birches. The foxgloves stood like slight maidens, with crowns of fair purple blossoms and long arms reaching for the star-strewn sky. The oaks were kings and the foxgloves their daughters.

Notice the wonderful richness of the sensory details, the analogies, and the wonderful animistic imagery in this passage. Notice also the remarkable clarity of the simple subject-verb-object structure of both the main and the relative clauses. When individuals with dyslexia learn to write well, this clear, direct, image-rich style quite often characterizes their work. The structural similarity with Anne Rice's highly lucid writing is apparent, as is Penny's remarkable literary potential.

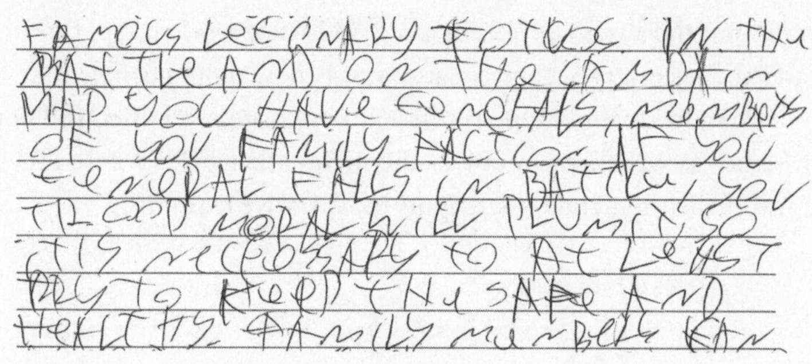

FIGURE 2

If you're like most people—including many teachers—you might wonder how a child with significant dyslexic challenges could have written this passage. Part of the answer is that it was written using a word processor with a spell-checking function. To show you how essential this technology is for Penny (and for many other children with dyslexia), let's compare the above passage with the sample of her spontaneous handwriting, as shown in figure 2. In this passage, Penny was writing about her favorite game. Notice that she's chosen to write it entirely in capital letters to eliminate reversal errors. Since parts of her writing can be difficult to read, we'll "translate" the passage here for you, with spelling errors retained:

> . . . FAMOUS LEGENARY TOTUOUS [ed. "tortoise"]. IN THE BATTLE AND ON THE CAMPAIN MAP YOU HAVE GENERALS, MEMBERS OF YOU FAMILY FACTION. IF YOU GENERAL FALLS IN BATTLE, YOU TROOP MORAL WILL PLUMIT, SO IT IS NECESSARY TO AT LEAST TRY TO KEEP THE SAFE AND HEALTHY. FAMILY MENBERS CAN . . .

It's hard for most teachers when faced with such handwriting to perceive the impressive literary talent lying underneath. But it really is important to look, because N-strengths are very common in students with dyslexia, and in our experience there are many more Pennys and Anne Rices and Blake

Charltons out there than anyone suspects. With appropriate support, strategies, and the necessary accommodations, it is now more possible than ever for dyslexic individuals with powerful N-strengths to reach their full potential, whether as writers or in any of the many other fields where they can make use of their remarkable strengths in Narrative reasoning.

PART VI

D-Strengths

Dynamic Reasoning

CHAPTER 20

The "D" Strengths in MIND

When she was a child struggling in school, Sarah Andrews's mother called her "my little underachiever." As you'll see, it's been a long time since anyone has called Sarah an underachiever.

Like many individuals with dyslexia, Sarah was a so-called late bloomer. And like many dyslexic individuals, Sarah's "blooming" was less like the gradual unfolding of a bud into a blossom than the astonishing transformation of a caterpillar into a butterfly.

Sarah was born into an academically accomplished family of teacher parents and honor student siblings, but right from the start she struggled with many academic skills, including spelling, math calculations and procedures (especially "showing her work"), and rote memory of all kinds. But her biggest challenge was learning to read. Sarah described for us some of her early reading problems: "The letters and print vibrated on the page. I saw the fibers in the wood pulp in the paper. I put my head so close to the page that my chin was right on it, and my teacher put a ruler under my chin to keep my head off the page."

Because of her difficulties, Sarah was unable to get through a first-grade reader until third or fourth grade. (To this day her reading remains agonizingly slow.) She had to read each sentence several times to understand it, and she had difficulty keeping her mind from wandering because each word "set off cascades of ideas and associations" that she felt she must "test and integrate."[1]

As Sarah recalls of her early years of school, "I was taking in tremendous amounts of information and starting to group it and sort it and arrange it, but there wasn't anything coming out—there wasn't a product." Sarah's mother—who also happened to be the English teacher at the small private school Sarah attended—was determined to remedy this lack of output, so she drilled Sarah on writing, particularly on writing essays. To the delight of both, Sarah found (like Anne Rice) that it was "easier to code than to decode."

Unfortunately, reading remained a problem. When Sarah reached high school she still "couldn't read a lick," and it finally caught up with her on the SAT. Although she'd received "decent" grades in school, her performance on the SAT was so poor that it astonished her teachers.[2] As Sarah recalls, "Someone finally asked if I finished the test . . . I said, 'I got about halfway through.'" The riddle of her poor performance was solved—almost.

Sarah was packed off to the remedial reading lab. Because she could pass all the phonics tests but just couldn't read fluently or retain what she'd read, the reading teacher improperly stated, "You don't have dyslexia. You're just lazy." Despite this misdiagnosis Sarah practiced diligently, and she improved her reading speed enough so that when she retook the SAT, she not only finished but doubled her score. Even with this improvement, though, Sarah still hadn't reached grade-level reading proficiency. She was, in her own words, "a bright high school senior now reading at the eighth-grade level."

After graduation, Sarah turned down admission to two top art schools and enrolled at Colorado College. She wasn't sure at first what she wanted to study, and like many college students with dyslexia, she struggled in the courses she was required to take during the first two years. She took a course in poetry to meet her English requirement—primarily because she thought it wouldn't require much reading—and received the first and only F of her career. To make up for those credits, Sarah took a course in creative writing. To her delight and surprise, she found that she not only enjoyed writing fiction but actually had a knack for storytelling. Although she couldn't "sustain the activity" of her stories for more than three or four pages, her teacher praised her stories as outstanding miniatures. Sarah didn't realize it then, but this newfound skill would play an important role in her life.

Another life-changing discovery came when Sarah took her required science course. She selected geology, almost on a whim, because the only scientist in her family, her aunt Lysbeth, was a geologist and, like Sarah, she was dyslexic.[3]

Sarah soon realized that in geology she'd discovered "the right playground for my mind. . . . I at last found teachers who perceived my talents, and I could learn from maps and illustrations rather than insurmountable texts. For the first time I was around a concentration of people who thought like I did, and I wasn't being snubbed as a weirdo."

Even among these like minds, Sarah was delighted to find that some of her talents were exceptional. "I was the best map interpreter—I was really quick at taking in graphical information holistically, seeing the patterns, understanding their meaning, and making interpretations from them. Being the best at something in class was a new experience for me, so I stuck with it." As she did, the self-doubt produced by the earlier labels of "lazy" and "underachiever" began to fade away. "By the time I noticed that I was not in fact lazy, I had earned both a B.A. and M.S. in geology."

For her first job as a geologist, Sarah went to work as a research scientist for the U.S. Geological Survey. She was assigned to study modern sand dunes in order to determine how gases and fluids could be removed from rocks that had been formed in prehistoric times from similar dunes. During this work, Sarah found that she was especially good at visualizing physical bodies in three dimensions and at imagining how processes would act on those bodies over time. These skills made her especially good at detecting analogies between modern dunes and ancient rocks and at predicting the structure and behavior of buried rock formations.

After leaving the USGS, Sarah went to work as an exploitation geologist (i.e., a geologist who specializes in finding ways to remove known oil deposits from the ground) for several oil and gas companies. Sarah's role with these companies was to improve oil and gas extraction from drilled wells by predicting how these substances would move through the surrounding rocks. Here, too, her spatial imagery and pattern-reading abilities proved invaluable. Sarah found that she was especially good at reading the "wire line log"—an immensely helpful but almost bafflingly complex visual readout of the physical characteristics of the rocks and fluids surrounding a well shaft. Sarah quickly learned how these "squiggles on a page" (which resemble the EEGs neurologists use to analyze brain activity) could predict the properties of surrounding fluids and rocks. Sarah found that she could transform these abstract squiggles into mental 3-D images: "I could visualize in time and space how the oil was going to move through the rock, even when it was fragmented and shattered."

Clearly, Sarah found a profession that seemed tailor-made to fit her mind. Geology drew heavily upon her strengths, placed little strain upon her weaknesses, and provided her with an endless string of fascinating puzzles to captivate her intellect.

The question we'll examine in the chapters ahead is, What *are* the strengths that so powerfully equipped Sarah for work in geology?

Dynamic Reasoning: The Power of Prediction

Sarah's "geological reasoning" abilities no doubt are due in part to her outstanding M-strengths. Sarah's powerful 3-D imagery system allows her to mentally visualize and manipulate full-color, lifelike imagery, which she finds incredibly useful for tasks like map reading, navigation, and remembering 3-D environments.

Yet if we look closely at the full range of the reasoning skills Sarah uses as a geologist, we can see that they're not entirely spatial. Geological reasoning requires more than simply visualizing and manipulating spatial images on spatial principles alone. It also requires the ability to *imagine* or *predict* how those images will change in response to *processes* that aren't entirely spatial in character, like erosion, earthquakes, sedimentation, and glaciation. These processes involve complex, dynamic, and variable blends of factors and are themselves often subject to larger processes, like climate variation or plate tectonics.

We call the reasoning skills that are needed to think well about such complex, variable, and dynamic systems Dynamic reasoning, or the D-strengths in MIND.

> D-strengths create the ability to accurately predict past or future states using episodic simulation. D-strengths are especially valuable for thinking about past or future states whose components are variable, incompletely known, or ambiguous, and for making practical, or "best-fit," predictions or working hypotheses in settings where precise answers aren't possible.

From one perspective, D-strengths can be seen as a subset of N-strengths: they're based primarily on episodic simulation, which is a component of N-strengths. Yet D-strengths are important enough, complex enough, and distinctive enough in their applications that we believe they deserve their own listing within the MIND strengths.

The key difference between N-strengths and D-strengths is the distinction between *creativity* in general and *creative prediction.* As we described in chapter 16, N-strengths include all the functions of the episodic construction system, each of which works by combining elements of past experience to construct lifelike narratives or "scenes." These scenes may "restage" actual past experience—in which case we call them *episodic memories*—or they may recombine elements of past personal experience in entirely new ways to form *creative simulations* or *works of imagination.* It's only when these creative constructions aim at predicting future events, reconstructing past events we didn't witness, or solving new problems, that we say they involve D-strengths. When the episodic simulation system is used to recombine elements of memory to entertain, persuade, paint an arresting picture, or form a compelling vision—but not to predict or reconstruct actual events or conditions—we refer to those actions as N-strengths but not D-strengths.

D-strengths aren't simply creative in a general or unconstrained way. They aim to predict—to simulate the world as it really exists, has existed, or will exist. D-strengths use episodic simulation to construct *true narratives*; they are N-strengths in their most practical mood, with their work clothes on and their sleeves rolled up. N-strengths may simply aim at being compelling. D-strengths must be *accurate*.

Let's compare the similarities and differences between D-strengths and N-strengths by seeing how Sarah employs them in her two very different—yet surprisingly similar—careers.

CHAPTER 21

The Advantages of D-Strengths

The first time we spoke with Sarah she quickly informed us, "I'm going to tell you everything as a story, because that's how I experience the world." The conversation that followed proved this assertion to be entirely true.

It's not surprising that Sarah has a highly narrative memory and reasoning style. As we discussed in part 5, many individuals with dyslexia do, and Sarah displays many of the features commonly seen in such individuals. For example, Sarah is a classic "family historian with a poor memory." As she told us, "I'm the family elephant: all my cousins come to me when they want to find out what happened when and where." Sarah also struggles with rote memory, which is common for individuals with dyslexia with this pattern: "I have no rote memory at all. I can only remember things if they fit into a structure." Typically, this structure is a story: "Stories are what I remember—they stick in my memory."

In fact, narrative is more than a reasoning style for Sarah. It's also a second career. Sarah Andrews is the author of ten highly regarded mystery novels that feature the exploits of professional geologist and amateur detective Em Hansen. In these novels, Em uses her skills as a geologist—and her prodigious powers of episodic simulation—to solve mysteries.

But the first problem Em solved was one of Sarah's. Sarah told us, "When I was in my thirties and working in the high-stress atmosphere of the oil business, I had trouble settling down to do my geology work if I'd witnessed an event that had a strong story to it. But I discovered that if I went ahead and wrote down the story or anecdote . . . I could squeegee it out of my mind and focus on the work I was supposed to be doing. Writing just seems to get the stories out of me."

Sarah described the mental mechanism she uses to construct her stories in a way that closely mirrors our description of episodic construction: "As time passed, the anecdotes gathered like lint in my mind, so I made fabric from it, and the bits of fabric needed to be rearranged in order to move tensions and troubles toward resolution. . . . [This fabric formed] a patchwork quilt of memories, in which I took various events and reorganized them into new events."

Sarah soon realized that before she could turn these "patchworks" into novels, she needed to figure out how to explain to nongeologists what geologists do and how they think; but before she could do that, she had to explain those things to herself. This preparation required two solid years of introspection, but it was worth the effort because it enabled Sarah to describe geology in all its complexity and wonder in a way that her readers could understand. It also led to many deep insights about "geological reasoning," which Sarah has described in several fascinating essays. One of these insights concerns the narrative nature of geology. Although the rocks and minerals that are geology's focus may initially seem "nonnarrative" and "impersonal," Sarah insists that geology is every bit as narrative as mystery writing because the rocks tell stories if we are tuned to listen.

Sarah isn't alone in this contention. Jack Horner, the world-famous paleontologist (and dyslexic) whom we met in chapter 11, has also identified a narrative element in geology, which he describes in his book *Dinosaurs under the Big Sky*: "Geology is the most important science for a dinosaur paleontologist to know because we find dinosaur skeletons in rocks. Our knowledge of geology helps us understand where to look, what to look for, and how old the fossils are. Geological information is essential because it helps paleontologists figure out what happened to animals, what may have killed the animals, and what happened to their remains after they died. Geology tells us the stories in the rocks."[1]

Narrative construction in geology differs from writing novels primarily in the constraints that reality places upon the construction process. While the novelist seeks to create an interesting and compelling scene that resembles something that *could* happen given particular hypothetical conditions—a scene, as Sarah told us, that "rings true"—the geologist tries to use available facts to predict precisely what the earth's past or future conditions actually were or will be. This requirement that constructed scenes accurately predict distinguishes D-strengths from N-strengths.

Dynamic Reasoning: How Elephants Become Prophets

The role that episodic simulation plays in geological reasoning can be seen in the description that Sarah gave us of her research on rock formation: "I took in everything I had ever observed and projected myself backward in time, seeing the landscape on which the sands had been deposited before they became rock."

This description reflects several of the features of episodic construction that we've already discussed. "Taking in everything" means forming memories through observation, so the components of these memories can be used later for episodic construction. "Projecting myself backward in time" means combining memory fragments through episodic simulation into mental scenes that "predict" what the past was like. In other words, Sarah's method of geological reasoning involves constructing mental images of past landscapes by recombining memories of personal observations of the current landscape, rather than reasoning in a logical, sequential, step-by-step fashion using abstract principles or verbal or mathematical models.

Sarah further described this process of construction when she wrote of thinkers like herself: "We are great sponges for *observed patterns*, both the concrete patterns of visual observations and the more abstract patterns of process and response. . . . Repeated patterns become ideas, and new patterns lead to new paradigms. . . . [We] can, using the barest shreds, 'see' through solid rock, back through time, and into future events."[2]

According to Sarah, this constructive process can combine even the "barest shreds" of observed patterns—whether patterns of physical objects, like rocks or sand dunes, or of processes affecting those objects, like wind or flooding—to build mental scenes in which the remote past, unobserved present, or distant future may be simulated. Importantly, this constructive process isn't limited to building "static" or "snapshot" scenes of single points in the past or the future but can create a continuous and interconnected series of scenes that allow the observer to "see through time" in a form very much like a time-lapse film.

This ability to create *a continuously connected series of scenes* is especially valuable for imagining and predicting the effects of *processes* that take place over long periods of time, like erosion, flooding, or the movement of the earth's crust along fault lines. Each of these processes occurs at its own unique rate, or along its own "time dimension." Episodic simulation is valuable for thinking about these processes because it can be used to predict their combined effects without losing sight of the impact each exerts. It allows geologists to "mentally experiment" by independently varying these processes, changing the rate or extent of their effects. That's why it's the ideal tool for "reading stories" in complex fields and for generating hypotheses, evaluating solutions or plans, or predicting the possible results of different actions.

Sarah points to why episodic simulation is especially valuable in situations that are changing, uncertain, or ambiguous in the following statement: "By working qualitatively we can mentally bridge gaps without having to plug in assumptions, and, as a result, it becomes possible to work with uncertainties, rather than simply overriding them."[3]

By "working qualitatively" Sarah means working with data whose form still resembles the original observations. These data consist of memory fragments that have been acquired from the original observations and can be used to construct scenes that simulate past, present, or future conditions in a form similar to the original observations.

We can get an idea of why this "qualitative" episodic simulation–based approach is so powerful by comparing it with its alternative: *abstract reasoning*. Abstract reasoning uses verbal, mathematical, or symbolic abstractions rather than episodic memory fragments. These abstractions have been created by *combining and converting* the original observations into generalizations that differ from those observations in form. These abstractions are stored in semantic memory as decontextualized facts. These abstract generalizations are often useful for reasoning about routine or typical cases, but less useful for thinking about unusual, unexpected, or unprecedented cases. This loss of usefulness in unusual settings occurs because the very process of creating generalizations means that the data must be averaged out, which causes the information from exceptional cases to become effectively diluted out by the more typical results.

We can highlight some of the benefits of using more "primary" or "separated" data that better reflect the differences between cases with several examples. Think first of a baseball team with thirty players. The team's batting average can be calculated by combining the personal averages of all the players. Let's say our team's average is 0.250, or one hit in every four at bats. Now, most of the individual players will have averages that are close to 0.250, but a few poor hitters may have averages around 0.100, and a few excellent ones may have averages near 0.350. If we want to predict future performance, the team average will be fairly good at predicting the batting success of the team as a whole. It will also be fairly good at predicting the success of the average hitters. But this generalized average will be poor at predicting the performances of the best and the worst hitters. We could much more accurately predict the performances of these "outliers" by using their past *individual* performances—that is, by retaining the data in a form that better reflects the original events and doesn't dilute the information relating to these special cases with the more typical data.

Here's a second example that also demonstrates how qualitative reasoning can be better in new or unprecedented situations. Let's say we want to predict how a particular batter will perform against a pitcher who he or she has never faced before. That player's overall batting average—which combines the results of batting against all pitchers—will be less useful than considering how that player has batted against pitchers with styles similar to the new pitcher. This pitcher-to-pitcher comparison is precisely the kind of "qualitative" reasoning process that Sarah described, and it provides information that abstract generalizations cannot.

These examples help illustrate some of the advantages that D-strengths (like episodic simulation and qualitative reasoning) have for working and making predictions in situations where the important information is changing, incompletely known, or ambiguous. By building mental scenes using the "raw data" of observed patterns as they exist in the real world, rather than using verbal or numerical abstractions, we can arrive at practical, best-fit solutions for difficult, unusual, or unprecedented problems without having to "assume away" ambiguities. This process of "using what fits" rather than relying entirely on abstract analysis or secondhand models is extremely powerful for solving practical problems.

CHAPTER 22

Trade-offs with D-Strengths

While D-strengths can provide tremendous advantages in dynamic situations, they can create drawbacks in other settings. One of the biggest drawbacks is a reduction in speed and efficiency.

So far we've spoken of Dynamic and Narrative reasoning as if they were largely active processes in which a person constructs, searches, sorts, and simulates at will. Often, however, this isn't the case.

When we ask individuals with dyslexia who rely heavily on Dynamic or Narrative reasoning to describe their reasoning methods to us, it's striking how many portray a kind of backward process in which the answers appear first, essentially fully formed, after which a more conscious process must be pursued to connect this answer with the initial conditions. Douglas Merrill's description of his problem-solving method is typical: "I usually begin by visualizing what I think the end stage should be, then I work backward. I can't exactly describe what I do because it feels more intuitive than traditional storytelling or deduction."

Sarah Andrews echoed this description when she wrote of herself and others like her: "Given a problem and an hour to solve it, we typically spend the first three minutes intuiting the answer, then spend the other fifty-seven backtracking . . . to check our results through data collection and deductive logic." According to Sarah, this intuitive approach "functions in leaps rather than by neatly ratcheting intervals" and is "less lineal than iterative or circular."[1]

This intuitive approach—used very heavily by individuals with dyslexia who excel in Dynamic and Narrative reasoning—can be very powerful, but it does present a problem: when viewed from the outside it can look an awful lot like goofing off. Sarah shared an example of this from her own life. One day at work she was standing by her office window staring serenely out at the mountains while trying to let her mind "ease itself around a problem." Her CFO walked by her door, looked in, and saw one of "his people" staring out the window, so he snapped at her to get back to work. Sarah calmly replied, "You work in your way, I'll work in mine. Now stop interrupting me." Sarah later wrote of this episode, "What this CFO didn't know was that staring into space is precisely how we work. It is our capacity to throw our brains into neutral and let connections assemble . . . that makes it possible for us to see connections that others can't. We relax into the work."[2]

This need for patient reflection can also create enormous problems at school, where time for reflection is in critically short supply. Try convincing a teacher that staring out the window is how you work best or that "getting busy" means you'll get less done. Yet this passive and reflective approach really is a valid problem-solving method, and there's plenty of scientific evidence to support its validity and effectiveness. In the research literature, this method of problem solving is referred to as *insight.*

Insight involves the sudden recognition that connections exist between elements of a problem. The classic historical example of insight is Archimedes shouting "Eureka!" and leaping from his bathtub as he suddenly realized how water displacement could be used to measure the volume of irregularly shaped objects.

Insight is at its most useful when step-by-step or analytical problem-solving is hampered by the ambiguity or incompleteness of the information—that is, in situations where D-strengths are required. Insight is also highly dependent upon the I-strengths we described in part 4. This is because insight depends on the same broad and "distant" *cognitive* connections between concepts and ideas that are characteristic of I-strengths. Because insight is so closely linked with D- and I-strengths, one would expect to find that individuals with dyslexia are especially good at insight-based problem solving; and in our experience that's precisely what we do find.

Although insight-based problem solving is very powerful, because much of its connection-making process takes place outside the person's conscious awareness, it can often seem second-rate, mystical, shoddy, or even slightly disreputable. But there's an observable neurological mechanism underlying insight that's been well worked out over the last decade by researchers.

One of the scientists who has contributed most to our understanding of insight is Dr. Mark Beeman, whom we met in chapter 4 when we discussed his work on hemispheric language functions. Dr. Beeman has been especially instrumental in showing that the process of insight involves several distinct phases.

In the first phase, the mind focuses actively upon the problem at hand and sets out the questions that need to be answered. This highly focused phase quickly gives way to a *relaxation* phase, where the mind loosens its focus and begins to wander. As Dr. Beeman has described this stage, "There's an overall quieting of the brain's processing because it's trying to calm everything down and wait for something to pop out." That "something" the brain is waiting on is the recognition of "distant or novel associations or relationships,"[3] which are just the kinds of connections that individuals with dyslexia typically excel at making. When a suitable connection is found, it results in the simultaneous activation of a broad cellular network that stretches all over the brain. This widespread electrical burst creates the subjective sensation of the eureka moment.

Notice how closely this insight mechanism links mental states like relaxation, reflection, and daydreaming with productive abilities like creativity, the ability to detect distant connections, and the ability to solve problems. Research by Dr. Demis Hassabis and others has shown that the brain circuits that become activated during daydreaming or mind wandering (termed the brain's "default network") are essentially identical with the episodic construction system. In other words, daydreaming consists of free-form and undirected scene construction, or the creative recombination of episodic memories. This is largely what *imagination* means. Small wonder there's such an extensive overlap between imagination and insight, daydreaming and problem solving—or between staring at the mountains and solving difficult geological problems.

Factors like emotional well-being and positive mood also seem to play an especially large role in supporting successful insights, and they work by enhancing the relaxation phase. This is why many great insights seem to occur in showers, baths, beds, or on beaches or when gazing out windows or staring emptily into space. Attempting to "force" or hurry insight will only inhibit it. This is one of insight's most confusing features: its success seems to vary almost inversely with effort, so that the deepest engagement requires a kind of deep disengagement. The harder you try to solve some problem using insight, the less likely you are to succeed. Insight is most likely to occur when the mind is wandering in a relaxed state rather than when hurrying toward a specific goal.

This isn't the first time in this book that we've stated that tight mental focus and attention can inhibit creative connections. Think back to our discussion in chapter 12 of latent inhibition, where we mentioned that tight mental focus and resistance to distraction are inversely correlated with creative achievement. In contrast, making distant, creative, insightful connections may be fostered by a slightly leaky attentional system, which allows ideas to mix.

Childhood may be the time in life above all others when nature favors us with the capacity to make creative and insightful connections. Dr. Beeman speculated to us on the value that the human species' unusually prolonged period of attentional immaturity may play in the development of creativity: "Perhaps there's some benefit in the delayed development of mental focus—maybe that's why humans in general develop so slowly. And perhaps those children who are developing more slowly in their attentional skills are developing more richly in certain aspects of creativity; and perhaps this extra creativity means that they'll actually develop in great ways if we don't mess them up too much in the meantime."

Dr. Beeman was clear about the kind of "messing up" he had in mind. "One big concern I have is with the use of stimulant medications used for ADHD [e.g., Ritalin, Concerta, Adderall, Vyvanse]. These drugs improve mental focus and resistance to distraction, but getting people to focus more may ultimately be bad for creative thinking. We may actually be inhibiting growth in areas like creativity and insight that are very useful—and that's something that we really should not do unless it seems absolutely necessary."

Rather than judging a child's development solely on qualities like speed, quantity, and focus during work, we should be monitoring their development of creativity, use of insight, and time spent in reflection. By failing to recognize the value of the slower but incredibly rich insight system, and instead placing all our emphasis on linear, rule-based, deductive thinking styles, we hinder the development of all children, but perhaps especially those who are the most creative and insight dependent.

One area where we often see insight-based problem-solvers suffer unnecessarily is in math. It's not uncommon to find phenomenal young mathematicians in our clinic who are unable to show—or in some cases even describe—the steps in their work, yet who get nearly every problem right. These students are solving problems through insight—matching the patterns of new problems with ones they've seen before, and sorting through their memory stores for answers that fit rather than employing step-by-step analytical reasoning. While it's important to gradually help these students learn to trace through the intervening steps, in the preteen years so long as they are able to consistently reach right answers they should be given credit for their understanding even if they fail to demonstrate all the steps in their work. As they age and their long-distance circuits become better insulated, their efficiency at moving between insight and analytical problem solving will improve, and they'll be better able to "reverse engineer" the intervening steps and to show their work. Unfortunately, we've seen some truly disastrous cases where profoundly gifted young mathematicians have lost their love for math simply because too much emphasis was placed on having them show their work when they weren't developmentally ready to do so.

It's important to recognize that certain individuals are simply predisposed to solving problems through insight rather than analysis. In our experience, this is true of many individuals with dyslexia. People whose reasoning is based primarily on insight can sometimes look unfocused, inefficient, "nonlinear," or slow to others who don't fully grasp the nature of the insight mechanism they're using, and they often have difficulty getting others to accept the results of their reasoning process if they can't "show their work." However, insight-based reasoning deserves far more respect than it receives. As teachers, parents, co-workers, and bosses, we need to be watchful for individuals who frequently reach the right results through insight, and when we find them we need to treat their different reasoning style with the seriousness it deserves. Not all staring out the window is productive reasoning, but quite a lot is; and it's important to understand that some people—including many of the most creative—really do need to "relax into their work."

CHAPTER 23

D-Strengths in Action

Things They Don't Teach You in School

We've seen how Dynamic reasoning can help in situations that are changing, uncertain, or ambiguous. Now let's look at some intriguing evidence that demonstrates how individuals with dyslexia as a group have achieved conspicuous success in one of the most changeable and uncertain environments of all: the world of business.

Let's begin by looking at one dyslexic entrepreneur who's shown a knack for building successful businesses. His name is Glenn Bailey.

Since Glenn began his first company at the age of seventeen, he's built many profitable businesses in a wide range of sectors including services, construction, and retail. Yet for all his success in the business world, Glenn rarely found success in school.

"My school career was dismal. I had a hyperactive mind, so my focus was just not there. My mind tended to wander a lot—usually in order to entertain myself. I'd be off in another world and thinking about other stuff.

"My mind is very visual: I can see anything in pictures, and I always visualize things. I can't help it. It's how I'm wired. So whatever you talk about, I'll see pictures in my head. Very vivid, colorful, lifelike pictures. They aren't still pictures. I can make them move. Reality, fiction, whatever. I really have to pull it back in to get focused. It was also a problem in the classroom because I'd sit there and imagine where I'd want to be, and what I'd want to do, and what I wanted to become, and I'd think happy thoughts, and I'd just be tuned out the whole time in class. I'd sit there nodding and smiling, but really I was like, 'What are you talking about?'

"I was also very curious, and strangely enough that became an issue at school. I'd ask questions like, 'Why's this?' or 'Why's that?' and that was treated like a problem. Teaching was regarded as a one-way street, but I really learned best through interaction.

"My biggest problem was in English. Being able to read is the 'face of intelligence' you present to society, and if you can't read, people just automatically assume that you're stupid. What happens to individuals with dyslexia in school is that reading becomes this big fifty-pound weight that just drags your whole body under. So I didn't have much confidence in any department of academics. I just thought I wasn't that bright. I called myself 'the Shadow' because I was just trying to get by day by day."

Glenn left school at age seventeen, when he found that it wasn't adding much to the strengths he intended to use in the real world. "The last day I ever went to school I found myself studying for English—which I didn't excel in or even like—during my math class. And I thought that was quite ironic, because I love math and I love numbers. They're very logical, and I can do math in my head quite quickly. Yet here I was studying for a course that I hate—English—in a math class so that I wasn't learning math. So that was it for me. I left school, and that became a huge trigger point in my life. I said, 'Look, I can't do anything about the last sixteen to seventeen years, but I can do a lot about the future. Where do I want to be in the next five, ten, fifteen, twenty, twenty-five years?'"

So Glenn set goals, and he began pursuing them. After opening and running his own ski shop for several years, Glenn spotted an opportunity to introduce bottled water to Vancouver. At the time he was twenty-three. Within ten years Glenn's company had thirty-two thousand accounts and annual revenues of $14 million. He was so successful in building the Canadian

Springs Water Company that he became known in his hometown of Vancouver as "the Water Boy." In 1996 Glenn sold his share of the business for $24 million and received a Business Development Bank of Canada Young Entrepreneur Award. Since then he's built many more successful businesses, including another water business, which provides on-site purification rather than delivery of bottled water.

When we asked Glenn to identify the keys to his success as a "serial entrepreneur," he mentioned his abilities to spot opportunities and develop a vision, but he also cited his ability to form relationships with other people: "For me—as for any dyslexic—it's about having the right people around you. Motivating and delegating is a massive part of what I do. You can't do everything by yourself. I rely on people a lot, and on their support. I also have amazing family—parents who always loved and believed in me absolutely—and friends, and teachers, and my wife is absolutely super. Everything I've ever done I owe to them."

Dyslexic Entrepreneurs: A True Growth Story

Glenn Bailey is a true success story, but he's far from alone in being a successful dyslexic entrepreneur. At the time of this writing, a Google search on the term "dyslexic entrepreneur" calls up over thirty-seven thousand links. Many of these links bring up lists of dyslexic entrepreneurs or biographies of particular success stories like Glenn. But a substantial number also link to the work of Dr. Julie Logan, who is professor of entrepreneurship at the Cass School of Business, City University, London. While Dr. Logan isn't dyslexic herself, for over a decade she has studied many dyslexic entrepreneurs and has published several widely quoted studies on her work. We spoke with Dr. Logan and asked how she first became interested in the special abilities of dyslexic entrepreneurs.

"I used to spend a lot of my time doing management training and strategic development for managers in large firms, then I moved over to working with entrepreneurs. I quickly noticed a difference. While many of the entrepreneurs were very good at presenting a clear and convincing picture of the business they were trying to build, they were very reluctant to get their ideas down on paper and to produce a written business plan. That was something I hadn't come across when I was working with corporate managers. A manager with a large company could typically produce a very good written strategic plan. So I found that rather strange, and I began to wonder why that was. That's really how I first got interested. After that, whenever I met entrepreneurs who were great at communicating visions but seemed reluctant to put their ideas down on paper, I started questioning them about how they'd done at school, and whether any of their children were dyslexic, and so forth. So that's how my interest in dyslexic entrepreneurs developed. That pattern kept coming up again and again."

Dr. Logan eventually conducted formal research on entrepreneurs, both in the United Kingdom and later in the United States. In the United Kingdom she found that the incidence of dyslexia among entrepreneurs was about twice the rate it was in the general population, and in the United States it was at least three times as high.[1] In the United States, fully 35 percent of the entrepreneurs she surveyed were dyslexic, while fewer than 1 percent of U.S. corporate managers were.

Dr. Logan found several key traits among dyslexic entrepreneurs. The first is a remarkable sense of vision for their businesses. "They've got a very clear idea of where they're going and what they're doing, and holding that end point in sight is a very powerful tool because it can be used to harness other people around that vision."

Second is a confident and persistent attitude. "Having got through and solved all the problems of schooling and coped with those challenges, they have a can-do approach that they bring to new situations. They not only know what they want to do, but they're also confident that it's going to work."

Third is the ability to ask for and engage the help of others. Dr. Logan found that most dyslexic entrepreneurs employed significantly larger staffs than nondyslexic entrepreneurs, and they were more likely to delegate operational tasks to their staffs while focusing their own attention on the overall vision and mission of their companies. "They know they're not particularly good at the details, so they surround themselves with people who are good at doing finance, or good at attention to detail, or whatever it happens to be; and unlike many entrepreneurs who won't delegate and will constantly interfere, they'll bring the best people around, and then they'll trust those people to do it. Many individuals with dyslexia have used the same sentence to describe this attitude to me: 'I employ the best people I can find, even if they're cleverer than me.' That comes out again and again."

The fourth trait is excellent oral communication, which they use to inspire their staffs. "Often they have very personal charismatic relationships with the people who work for them, and even when they have enormous empires they manage to somehow create those same sorts of relationships. The employees are energized by them. A good example of this is how the staffs of both British Airways and Virgin Atlantic were both going to go on strike. They both wanted pay rises. But Richard Branson [the dyslexic CEO of Virgin Atlantic] went to the staff conference, spoke to the staff, and explained why he couldn't give them a pay rise; and everybody went back to work and carried on." By contrast, the CEO of British Airways didn't use that approach, and their labor standoff continued.[2]

A fifth and final strength, which Dr. Logan has repeatedly observed but has not yet confirmed with her research, is that "many of these successful entrepreneurs use their intuition a lot. For example, I've been talking recently to a very successful dyslexic entrepreneur, and he told me that he never does formal market research. He just goes and he'll stand next to a store that he's thinking of purchasing, and he looks at the footfall, and that sort of thing. It's certainly more of a right-brain approach than a logical reasoning approach."

In short, the skills that Dr. Logan has identified in these dyslexic entrepreneurs consist to a surprising degree of the kinds of D- and N-strengths we've covered in this part and the preceding one: Dynamic reasoning to "read" future opportunities, establish end points, and solve problems, and Narrative reasoning to convey the vision to others, inspire them, and persuade them to join in. These are the kinds of skills needed to operate in any environment that is changing, uncertain, or incompletely known.

CHAPTER 24

Key Points about D-Strengths

In these chapters, we've discussed the D-strengths that many individuals with dyslexia exhibit. Key points to remember about Dynamic reasoning include:

- Dynamic reasoning is the ability to "read" patterns in the real world that allow us to reconstruct past events we haven't witnessed, predict likely future events, or simulate and preview plausible outcomes of inventions or various courses of action.
- It is especially valuable in situations where all the relevant variables are incompletely known, changing, or ambiguous.
- Its particular power lies in the fact that it is based on matching patterns that are similar in form to the original observations, rather than on abstract generalizations.
- Dynamic reasoning often employs insight-based processing, which is powerful but often slow, can appear passive, and may result in difficulty explaining intervening steps.
- Its effectiveness in exceptional situations is enhanced by strong abilities in pattern separation, which increase the available memory patterns,

and by a highly interconnected brain circuitry that favors the ability to link to more distant and unusual patterns.

- Individuals with dyslexia who possess prominent D-strengths often thrive in precisely the kinds of rapidly changing and ambiguous settings that others find the most difficult and confusing.

The Power of Prediction: Vince Flynn

In closing, let's look at an individual who shows remarkable D-strengths—novelist Vince Flynn. Over the last decade, Vince's twelve counterterrorism-themed novels have sold over 15 million copies, making him one of the top-selling novelists in the world today. Vince's stories are renowned for their intricate plotting and their surprising twists and turns; but nothing he's written can approach his own life story for unexpected plot developments.

Vince was diagnosed with dyslexia in the second grade, when he was given the label "SLBP" ("slow learning behavior problem") and placed in a special ed class. Vince recalls his problems with reading, writing, and spelling as "nearly incapacitating," and he got by primarily by becoming a master of classroom survival tactics. As he once recalled, "[I] knew how to figure out the game and was always respectful to my teachers and tried hard. As long as you did that, it didn't matter how poorly you did on tests; they were going to pass you. . . . You participate in class discussions and the teacher says, 'This kid gets it; he just doesn't test well.'"[1]

Vince graduated from high school with a C+ average, then enrolled at the University of St. Thomas in St. Paul, Minnesota. There his lack of reading and writing skills immediately began to cause problems. After his first semester he was placed on academic probation. He took two English classes and got a C– in both. He barely scraped by until his junior year, when two things happened that changed his life forever. The first was an event that Vince told us he still recalls as one of the most humiliating of his life.

"I was taking a class pass/fail, and I handed in a paper to the teacher, but I wasn't there the next class to pick it up. So I had a so-called friend pick up the paper, and later that day I went to meet him for lunch, and there was my paper sitting right in the middle of the table with eight guys gathered grinning around it. And at the top of the page there's a big red 'F,' and toward the bottom the professor had written, 'I don't know how you ever got into college, and I don't know how you're ever going to graduate, but this is the worst paper I have ever read in all my years of teaching.' I was so embarrassed that I said to myself, 'This is it, I can't keep going like this. I've got to face this head-on.'"

"Right after that, Al McGuire [the Hall of Fame college basketball coach and TV commentator] came through St. Thomas on a speaking tour, and I was thunderstruck by what he had to say. He talked about how he had grown up dyslexic, and he didn't know how to read or write; but he was a standout basketball player, and he b.s.'d his way through high school and college with never better than a C average. After playing in the NBA for a few years, he began coaching the Marquette Warriors, and in his last season there they made it to the NCAA championship game. About fifteen minutes before tip-off the scorekeeper came up and handed McGuire the scorebook and said, 'Coach, I need your starting lineup.' And McGuire broke into a cold sweat because he had no idea how to spell his players' names. So he panicked and said, 'I can't—I have an emergency,' and he ran back to the locker room and locked himself in a stall, and he started praying, 'Dear God, if you let me win this game, I'll go back to night school and learn how to read and write.' So Marquette won, and Al McGuire retired from coaching and went back to school.

"Now, what I learned from this story was that the longer I put this off, the worse and more embarrassing it would become. So I really got serious because I wanted to write without being embarrassed and be something more than a functional illiterate.

"So I went out and bought two books, *The Fundamentals of English* and *How to Spell Five Words a Day*, and I started going to the library every day and really working through them. I also started reading everything I could get my hands on, because I knew that was the only way to get better."

Vince started by reading *Trinity* by Leon Uris. Although he struggled during the first hundred pages, he was soon hooked, and he stopped worrying if he couldn't decipher all the words. He found that his mind could "fill in the blanks."

After graduating from college, Vince took several sales jobs while trying to enter Marine aviator school; however, the several concussions he'd suffered playing football disqualified him. He spent several years pursuing a medical waiver that would allow him to fly, but when he hit twenty-seven and was no longer eligible for officer training, he began searching for another direction for his life. "I remember saying to myself there's no way I'm going to spend the rest of my life sitting in a cubicle," so he started working on the manuscript that would become his first novel, *Term Limits.* Soon after, he "burned his boats," quitting his job and telling his friends and family he was going to make his living as a novelist.

We asked Vince what convinced him he could make a living as a novelist when only a few years before he'd been told that his writing set new standards for ineptitude? He responded that it was the confidence that came from finding that his mind worked a lot like the novelists he loved. He said that when he read or watched movies, "I always knew what was going to happen next. A lot of times I could tell almost from the first chapter how it was going to end." That experience made him think that if he could predict the plots of other people's novels, maybe he could create his own.

Vince's suspicions were on solid ground. As we've discussed, prediction and narrative are closely related mental skills. However, Vince's ability to predict plot twists might not have been enough to make him risk his livelihood if he hadn't also seen other signs that he had special skill in dealing with complex patterns. One was a surprising talent he'd shown as a child: "I was a naturally gifted chess player. It was a weird, weird, weird deal: even though I was failing in school, I could always just see in advance how the game would unfold. So I used to get driven around by my parents and dropped off at the houses of some of the best chess players in the Twin Cities. One year I actually finished fourth place in state. But I never told my friends at school because I was embarrassed about it."

Another hint came in college. "While I was struggling in my other classes, I ended up taking macroeconomics, and for the first time in my life I found myself sitting in a classroom being one of the only people who knew what was going on. A lot of the kids who were gifted in English and math were just scratching their heads because it didn't make sense to them—there were just too many variables—and suddenly I'm the one who's saying, 'How do you guys not understand this stuff? This is easy!' It just made sense to me."

Vince's powers of prediction were clearly revealed in his writing. After his first novel got him a contract with a major publisher, Vince chose for his second a topic that in 1998 most people were still largely unaware of but which struck him as the most important national security issue of the time: Islamic radical fundamentalism. Vince's next three novels all focused on this threat, and all were published before the tragic events occurred that brought this danger to everyone's consciousness. "Prior to 9/11 I wondered, 'Why isn't anyone else scared about this?' To me it was just obvious that this was a disaster waiting to happen."

Vince's ability to predict both the headlines and the behind-the-scenes action is so startling that one of his books, *Memorial Day*, attracted a security review by officials at the Department of Energy because they were certain he must have been fed classified information from an inside source. But Vince says the realism of his stories comes entirely from his ability to predict using freely available information. "I can honestly tell you that I've never had an active-duty person with the CIA or the Secret Service or NSA or FBI or armed services—or anywhere—give me classified information. I just take the information that's available in the public domain, and then I fill in the blanks. I have a way of connecting those dots."

People outside the government have also noticed Vince's predictive powers, and some of them are so impressed that they're ready to stake money on them. "I was recently asked to go on the board of directors for a hedge fund. These guys said, 'We think that you have a good way of strategic thinking,' and they don't know why, and they don't know a thing about dyslexia, but they said, 'We'd really like to get you on this board so you can do some strategic thinking for us.'"

Vince has also tried to convince others of the predictive power he believes many individuals with dyslexia possess. "I have a friend who's on the boards of several charitable foundations, and I'm always telling him, 'You know, every one of those associations needs at least one dyslexic on the board.' When he asks me why, I tell him, 'We just see patterns in advance. We could do you a lot of good.'"

Vince attributes this dyslexic predictive power to both nature and nurture. "Dyslexics are wired a little differently, and that probably makes us a little more creative innately, but there's also the impact of experience. School's like a wall, and for everybody else there's a ladder there, and they just get up on that ladder and climb over the wall. But for whatever reason, dyslexics don't know how to climb that ladder, so we've got to figure out another way to get past that wall. We've got to dig a hole under it, or find a rope to build a rope ladder, or find some other way around it. So we're constantly trying to solve a problem, and I think that's one reason why so many dyslexics become inventors and creators—because they're constantly looking for ways to beat that system, or improve that system, or change it so it makes sense to them. This develops their skills in forecasting as well. So many things in life are like an algebra equation where you're given four 'knowns' and three 'unknowns,' and you've got to solve that problem. Dyslexics are forced to do that so much in their everyday lives that it helps us become really good at solving complex problems."

After listening to Vince list all the benefits he sees to being dyslexic, we ventured to ask him, "So you really don't find it surprising that we've titled our book *The Dyslexic Advantage*, do you?"

Vince responded with a hearty laugh. "Hear, hear! You're correct. I most certainly don't!"

PART VII

Putting the Dyslexic Advantage to Use

CHAPTER 25

Reading

So far, in considering the many advantages that can accompany a dyslexic processing style, we've seen that:

- Dyslexia isn't simply a reading impairment but a reflection of a different pattern of brain organization and information processing that creates strengths as well as challenges.
- Dyslexia-associated strengths and challenges are inextricably connected, like home runs and strikeouts in baseball, and dyslexic challenges are best understood as trade-offs made in pursuit of other, larger cognitive gains.
- Individuals with dyslexia often show strengths in big-picture, holistic, or top-down processing, though they may struggle with fine-detail processing.
- Many individuals with dyslexia show strengths in Material reasoning, or the ability to mentally create and manipulate an interconnected series of three-dimensional spatial perspectives.
- Many individuals with dyslexia show strengths in Interconnected reasoning, or the ability to perceive more distant or unusual connections, to reason using interdisciplinary approaches, or to detect context and gist.

- Many individuals with dyslexia excel in Narrative reasoning, or the ability to perceive information as mental "scenes" that they construct from fragments of past personal experience (episodic memory).
- Many individuals with dyslexia show strengths in Dynamic reasoning, or the ability to accurately reconstruct past events that they didn't witness or to predict future states, often using insight-based reasoning and "episodic simulation," particularly in conditions that are changing, ambiguous, or incompletely known, and where "qualitative" practical solutions are required.

These findings have important implications for how we understand, educate, and employ individuals with dyslexia. In these final chapters, we'll explore these implications. Let's begin by looking at the learning function most closely associated with dyslexia: reading.

Becoming a Skilled Reader

Skilled reading requires three abilities: the ability to sound out words (i.e., decoding); the ability to read quickly and accurately (i.e., fluency); and the ability to understand what you read (i.e., comprehension). While individuals with dyslexia can struggle with any or all of these skills, their dyslexic advantages can also help them master these abilities.

Decoding Words

Decoding or "sounding out" unfamiliar written words depends on two key abilities:

- Accurately *identifying* all the component sounds in words
- Mastering the rules of *phonics* that describe how letters can be used to represent these component sounds

As we discussed in chapter 3, these abilities rely on the brain's phonological processing and procedural (or rule-based) learning systems. The phonological processing system works by first splitting incoming words into their component sounds (a process known as *sound segmentation*), then distinguishing (or *discriminating*) these sounds from one another. Most individuals with dyslexia struggle with one or both of these processes.

Problems with these processes can show up in a variety of tasks besides decoding. Individuals with dyslexia who have difficulty with sound segmentation will also often struggle to identify the component sounds in words. For example, they may have difficulty determining that "cup" has three sounds (*c-u-p*) rather than two (*cu-p*). They may also struggle to perform sound-switching tasks like Pig Latin (e.g., *ig-pay atin-lay*), or changing the *b* in *bat* to *h* to see what word is formed. With regard to sound discrimination, individuals with dyslexia who have difficulty distinguishing word sounds will often mispronounce, mishear, or misspell words. They may substitute or mistake similar-sounding words or make subtle soundalike mistakes in word pronunciations or spelling (e.g., substituting *t/d*, *m/n*, *p/b*, *a/o*, or *i/e*, or omitting word sounds like *pah-corn*).

Skills like sound segmentation and discrimination (which are referred to as *phonemic awareness* skills) aren't entirely inborn but must be learned. This learning takes place largely during the first two years of life and depends heavily on both fine-detail processing and implicit learning (that is, learning through observation and imitation, rather than *explicitly* learning the rules, as described in chapter 3). As we've mentioned, many individuals with dyslexia struggle with both fine-detail processing and implicit learning, so it shouldn't be surprising that they often have difficulty learning to accurately distinguish the full range of word sounds.

The good news is that the brain's sound processing system isn't fixed but highly reprogrammable. Brains with weak segmenting or discriminating skills can often be retrained with phonics instruction based on the Orton-Gillingham method. Individuals with especially severe sound discrimination difficulties should usually begin with an instructional technique that specifically improves the ability to distinguish word sounds, like Lindamood-Bell's LiPS, or computer-based auditory training programs such as Earobics or Fast ForWord. For all other individuals with dyslexia, an approach to phonics training should be chosen based on the individual's personal strengths, weaknesses, and interests, because these will together determine what kinds of information he or she remembers best.

While each individual is unique, the common dyslexic brain characteristics we've discussed in earlier chapters can also provide important clues about the kind of instruction that most dyslexic individuals will find beneficial. For example, since most individuals with dyslexia favor episodic over semantic memory, most will remember information about things they've *experienced* (or imagined as scene-based experiences) better than abstract or noncontextual facts. Individuals with dyslexia will also remember information better if they find it interesting and if they can place it into a larger framework of knowledge or understand its big-picture function or purpose.

An individual's MIND strengths can also help predict which training methods they will find the most effective. For example, individuals with prominent M-strengths usually benefit from methods that engage their strengths in spatial imagery. These typically involve various forms of visual, positional, or movement-based imagery. Finding a method that stresses the particular form of spatial imagery that an individual excels in (e.g., kinesthetic, visual) can greatly increase the likelihood of success.[1]

Individuals with impressive I-strengths typically learn well using methods that engage their ability to see interconnections. Such instruction often builds associations or analogies between new information and subjects they already know or are interested in.[2] These learners also tend to enjoy multisensory or multiframework approaches that present the same information in different ways. They frequently enjoy discussing how the approaches they're using work, as this engages their strengths in gist, cause and effect, and contextual thinking.

Students with especially strong N- or D-strengths often benefit from approaches that stress examples and cases, rather than just rules and definitions. Approaches that embed information in stories or events are also more memorable for these students, as are interactive approaches where learning takes place through conversation or interaction with an instructor (e.g., using discussion, dramatization, or game play).

In short, by carefully considering the dyslexic individual's particular strengths, interests, and challenges, you can more effectively match dyslexic students with an appropriate method of instruction. A complete discussion of the available methods is beyond our scope in this chapter, but additional information is provided on our Dyslexic Advantage website and in our book *The Mislabeled Child*.[3]

It's important to recognize that most students with significant dyslexic challenges will require additional reading instruction outside of school. Training should begin as soon as challenges are recognized, though it's never too late to start. For children with a strong family history of dyslexia, phonics instruction should begin as soon as they show an interest in learning to read, especially if they show any difficulties like slow speech acquisition, poor sound perception, misspeaking or mishearing, mispronouncing rhymes, or slowness learning the alphabet or letter sounds. With early attention, mild reading problems may be avoided altogether, and more significant challenges can be lessened in severity.

Families that are sufficiently dedicated can sometimes provide phonics training on their own, using commercially available materials; but in many cases getting help from a certified reading instructor, speech-language pathologist, or reading instructional center is a good idea. If a tutor is desired, the state branches of the International Dyslexia Association keep helpful lists of providers who are well trained in Orton-Gillingham approaches.

Also, don't fall into the trap of thinking that no additional reading training is required once a child's understanding of phonics reaches age- or grade-level proficiency. Children who can't yet apply their phonetic knowledge quickly and automatically (or *fluently*, as we'll discuss in a moment) usually struggle in upper elementary or middle school when they're asked to learn from texts or to decode new or unfamiliar words.

It's often particularly difficult to deal appropriately with older students who've already begun to read with reasonably good comprehension but who continue to show difficulties with spelling or decoding. When their language skills are strong, students with dyslexia can often learn to read silently with good comprehension, even when they remain poor at sounding out unfamiliar words (decoding) and at spelling (encoding). Often their problems with word identification escape notice until they reach high school or even college, when they begin to find words in their textbooks they've never heard spoken before, so they can't use contextual clues to guess the word's definition. At that point their decoding difficulties can create many practical problems. We've called the problem these students face *stealth dyslexia* because it so often evades the "radar" of detection, and it can sometimes be difficult to motivate these students to work hard on phonics instruction if they feel that their reading is "good enough." However, if they can be motivated to cooperate, the benefits they receive in improved reading and writing are usually well worth the effort of additional phonics instruction, particularly if they plan to go on to college.

Fluency

To read quickly and accurately enough to meet the demands of the upper grades, college, and the job market, individuals with dyslexia need to master more than phonetic decoding. They must also master a second key component of skilled reading: *reading fluency*. Fluency consists of both reading speed and accuracy, and it's acquired through extensive reading practice.

Reading fluency can be broken into several aspects, each of which can be developed by different types of practice. Before we discuss these fluency-building techniques, let's look at four key principles that should underlie any type of fluency practice.

The first key principle is interest. *Individuals with dyslexia learn best when their interest is engaged.* This phrase should be carved into the wall of every classroom. A state of heightened interest is often the *only* condition under which individuals with dyslexia can engage texts deeply enough to make progress with reading skills. That's why the first step in any form of reading practice should be to find something that the student *wants to read*. A book, magazine, website, comic strip, or anything else that truly engages the student's interest will always be better—even if it seems a little advanced for the student's current skill level—than something that seems more appropriate in difficulty but fails to capture the student's attention.

Interest is also important for selecting the right reading materials because a student's background knowledge can provide context and vocabulary that help the student "fill in the blanks" and identify difficult words. Students with strong M-strengths often enjoy books and magazines about physics, chemistry, engineering, inventions, mechanics, computers, airplanes, architecture, landscaping, automobiles, design, fashion, or other topics that engage their spatial interests. Students with impressive I-strengths often enjoy reading books that feature humor, interesting analogies or metaphors (like poetry, mythology, or fables), and multidisciplinary or big-picture approaches to complex topics like the environment, military or world history, or psychology. Students with strong N-strengths may enjoy books with a strong narrative element, like stories, fables, myths, histories, or biographies. Students with strong D-strengths are often drawn at younger ages to books of fantasy, science fiction, or mythology; and when older to books of mythology or to historical fiction that deals with imaginary or past worlds; or to books about business, economics, and finance; or to magazines that feature articles on entrepreneurship and technology. Of course, the individual student is typically the best source of information about what he or she finds interesting.

The second key principle is to take advantage of the top-down, big-picture, contextual processing strengths that many dyslexic students possess. One of the best ways to do this is by equipping students with a general idea of what they'll be reading *before* they begin to read. This practice has been consistently shown to improve reading fluency and to increase the pace of learning. There are several ways to perform this pre-equipping. First, the student can listen to the passage he or she will attempt as it's read aloud by a tutor or a recording. Second, the student can read or listen to a summary of the passage (e.g., from SparkNotes or CliffsNotes or a tutor). Third, reading books that are part of a series pre-equips students because such books usually contain familiar stories, words, characters, and contexts. Fourth, for longer stories the student can watch a film version before reading the book. As a variation of this approach, we've found that many students enjoy watching movies, then reading the script for the same movie, which can often be found on websites like the Internet Movie Script Database (www.imsdb.com).

The third key principle is to make sure before starting that the student knows all the words he or she will encounter in the text. Taking time in advance to point out difficult words and to make sure those words are understood and can be recognized will make fluency practice more effective.

The fourth key principle for fluency practice is to make sure that the author's sentence structures are easy for the student to understand. Many students with dyslexia (especially younger ones) have working memory challenges, which make comprehending longer and more complex sentences difficult for them. For these students, an author's choice of words, sentence lengths and structures, and topics can dramatically affect their comprehension and fluency. Students can usually tell whether a particular author "fits" their style of thinking well within a page or two. Authors who stick their subject right up front and avoid complex clauses and passive constructions are generally easier for dyslexic students to grasp.

With these key principles in mind, let's look at three different types of fluency-building practice.

Practice That Builds Quick and Accurate Sight Word Recognition. The first and most basic type of reading fluency practice focuses on building strength and accuracy in decoding and identifying individual words. To read fluently, a reader must be able to decode quickly and automatically and to recognize many common words by sight.

Although many students hate to hear this, for building these skills there's really no substitute for oral reading practice. Oral reading forces readers to accurately identify each word in a passage, and this requirement simply can't be enforced with any other type of reading.

For beginning readers, the method known as *guided repeated oral reading* has been shown to be the most effective. In this method, a "guide" first reads a passage of appropriate difficulty aloud as the student reads along silently. When the guide finishes reading, the student reads the same passage aloud. This passage is practiced daily until perfect fluency is achieved. For beginning readers one paragraph may be enough, but passages should get longer as the reader progresses. Eventually the reader may be skilled enough to skip the first reading and may instead simply read passages aloud while a guide or tutor follows along to check for accuracy and to point out mistakes.

When individuals with dyslexia read well enough to accurately identify most words in the passages they read, they can focus on faster silent reading. They can also practice building their speed by reading along silently while listening to a recorded book.

Beginning readers can also build sight word recognition by practicing with flash cards of some of the most common words. Lists of these words are posted on many websites and can be easily found by using the search term "Dolch words."

Practice That Builds Fluent Word-Level Problem-Solving Skills. Practice reading silently and independently can also help students build the kind of top-down problem-solving skills that allow them to guess the meanings of words they can't easily decode. For practice of this type, books or magazines with lots of pictures can be useful, as can being pre-equipped with an outline or summary of the passage to be read. Students should be encouraged to read quickly and to avoid getting bogged down on words they can't immediately identify. They should be told to read primarily for context and overall comprehension, and simply fill in the blanks as they move along. This kind of practice isn't sufficient by itself to build a truly skilled reader, but it does encourage stamina, interest in reading, sight word recognition, problem solving, and top-down contextual reasoning.

Practice That Builds Speed. Reading is like riding a bicycle: you need to move forward at least a certain speed to make it work. Many individuals with dyslexia have learned to decode individual words reasonably accurately but still read so slowly or laboriously that they can't absorb a coherent message from the sentences and paragraphs they're reading. Often individuals who decode relatively well but still read slowly aren't identified as dyslexic. Instead they get noticed for underperformance, inattention, lack of persistence with reading, or their tendency to avoid reading altogether.

Students who read accurately but slowly should practice reading along with a guide or with recorded books to help build up speed. Just as with exercise, where a treadmill that forces you to go its speed can keep you moving faster than you might choose to go on your own, reading along with a recorded book can promote faster reading. Newer electronic devices—including the text playback systems on most computers and e-readers—can often be adjusted to different playback speeds, so students can gradually increase their speed of reading.

Fluent reading also requires a well-functioning visual system, which is something many persistently slow readers lack. There's considerable controversy in the reading community about the role of vision and visual interventions in dyslexic reading challenges. We reviewed this controversy in detail in our book *The Mislabeled Child*, and as we wrote there, there really does seem to be a subset of individuals with dyslexia whose inadequate visual skills delay their reading progress and who can benefit from visual evaluation and treatment.[4]

These individuals in some ways resemble those we described above who have difficulty discriminating word sounds. While standard phonics training will eventually help most of the latter individuals improve their auditory discrimination skills, the speed with which they develop these skills often accelerates—in some cases greatly—if they receive computer-based auditory training. Similarly, visual training can accelerate progress for readers with severe visual symptoms. While reading practice by itself eventually improves visual functions in many (but not all) individuals, visual treatments can enhance the rate of reading development for people with severe problems with eye movement control and focusing. For some, the differences can be dramatic and can prevent prolonged underperformance or often uncomfortable symptoms like eyestrain or headache while reading; visual wobbling (where the letters seem to move); eye tearing; doubling of visual images; skipping lines or frequently losing place while reading; or behaviors like squinting, tilting the head, closing one eye, or putting the head very close to the page. Individuals with dyslexia who often experience visual symptoms while reading or doing other forms of fine-detail close-up work deserve a thorough visual evaluation. The appropriate specialist to perform this exam is a developmental optometrist who has specialty training in the kinds of functional eye skills that allow the eyes to work well for fine-detail work.

These specialists will usually have the letters F.C.O.V.D. in addition to O.D. after their name, and many can be located at www.covd.org.

Comprehension

To *comprehend* written materials, a reader must understand not only the meanings of the individual words used but also how those words relate to each other at the sentence level, and how higher-level language features like literary style, genre, implied meanings, and metaphorical language affect passage meaning. The reader must also have sufficient working memory capacity to keep all this information in mind during processing and must read fluently enough to register all the information before the memory tracing fades away.

It's important to recognize that not all problems with reading comprehension are caused by dyslexia. Students who read fluently and decode well but still comprehend poorly generally have other issues with attention or language. In contrast, students with dyslexia will generally show problems with decoding and fluency but will understand texts much better when hearing them read than when reading themselves. This difference helps pinpoint the source of their comprehension problems to the reading process itself.

Fortunately, once problems with decoding and fluency have been solved or bypassed through the use of recorded texts, the MIND strengths often make individuals with dyslexia especially good at comprehending texts. I-strengths often aid in recognizing associations (like symbolism, analogy, metaphor, irony, humor, or relationships of correlation or cause and effect), in using different perspectives or points of view to analyze texts, in comparing different texts to each other, and in identifying big-picture elements like gist and context. N-strengths can help dyslexic readers create scene-based imagery and keep track of the narrative threads connecting different parts of texts. D-strengths can help individuals with dyslexia think ahead as they read, which can make them very active, imaginative, and analytical readers. People are often surprised when we tell them that many of our older students choose to major in subjects like English literature, comparative literature, or history, fields that require a lot of reading, but when problems accessing textual information are solved, dyslexic students often become ideally suited to these subjects. In the next section, we'll discuss ways of increasing access to textual information for all students with dyslexia.

The Point of Reading: Using Technology to Increase Accessibility

The fundamental point of reading is to gain access to written information. Fortunately, with advances in technology that allow verbal information to be stored and transmitted in many ways, there's no longer any reason why individuals with dyslexia should lack access to information of any kind.

Civil rights advocate Ben Foss is a great example of how viewing dyslexic challenges as an issue of information access rather than as reading issues per se can lead to creative approaches at school and work. Ben is currently executive director of Disability Rights Advocates, a national civil rights organization that seeks equality and opportunity for people with disabilities. But until recently he worked as director of access technology at Intel's Digital Health Group, where he supervised the development of assistive technologies for individuals with disabilities. Ben's final project was the Intel Reader, a portable device small enough to carry in a purse or backpack that combines a digital camera with a text-to-speech reader. It can be used to read aloud, at up to five times normal speed, any kind of printed text that's in a location where it can be digitally photographed—whether from books, magazines, package labels, or signs on walls. The idea for the Intel Reader first started with Ben, and his team ultimately received two U.S. patents for technology related to its design.

Ben's interest in making texts more accessible to dyslexic readers came from his own experience as an individual with dyslexia. Learning to read was so hard for Ben that he abandoned it—in the conventional sense. As he told us, "You want to insult me really fast: tell me you can teach me to read. Because you can't. I've done all the remediations there are, and they just didn't take."

If Ben couldn't read, how was he able to earn a bachelor's degree from the prestigious Wesleyan University, a master's from the University of Edinburgh, and a combined J.D./M.B.A. from Stanford? Ben argues that the key to his success was his use of assistive technologies and other educational accommodations. We often describe accommodations as "interventions that get you out of unproductive activities and into productive ones." Ben adds the following helpful description: "Accommodations are a ramp for a wheelchair. They're a modification to a process that is still true to the end goal: the test is still the test, and the knowledge is still the knowledge, but accommodations provide a different way to gain access to it. I found that metaphor of a ramp to be a really powerful one."

One of the accommodations Ben found most helpful was recorded books. Like many individuals with dyslexia, Ben's primary access to recorded books was through the nonprofit organization RFB&D (www.rfbd.org), whose recordings he found to be invaluable. However, when he reached the even more demanding environment of professional school, he found that conventional recordings were no longer adequate. "I went to Stanford for a combined law and business school program, and they had an outstanding student disabilities office with everything for dyslexics. So I had a talking computer, and I had books on tape, but I found the latter to be problematic because if you're listening to text being read out loud at a normal speaking pace, you're actually listening at about one-third the speed that most people can read—and that wasn't good enough for me to keep up with my classmates." After careful searching, Ben found a solution that had been adopted by many visually impaired people: "I moved over to digital text."

By digital text, Ben means text that's been encoded in digital form so it can be read aloud by a text-to-speech program on a computer or other electronic device. Anything you type into a word processor becomes digital text, and according to Ben, the big advantage of digital text is speed. "You can listen to digital text much faster than analog text recordings—in some cases up to ten times the spoken rate, and over time you can train yourself to be comfortable with ultrafast digital text as a way of getting access to information. That's how people with visual impairments process text, and I learned that model."

Right off the bat, Ben was able to listen to digital text at two to three times the rate of spoken text. This was already a huge advantage, but over the next five years he became an even faster listener. "The ability to listen very quickly is built through a series of microsteps with a few significant turning points, but it's definitely a process. It's not like you take a pill and immediately you're good." Ben mentioned that one essential key to maximizing listening speed is to try many different electron voices to find the one that seems easiest to listen to.

Ben found digital texts to be a fully adequate approach to learning. "I never touched a book in law school. I learned to abandon reading. That was the only approach that allowed me to keep up with all the assignments. I chose a totally different road than most students, because for me going through the book road was like inching over an unpaved road, while going through my alternative formats was like speeding down the autobahn."

Another person we spoke with who has found accelerated listening to be a big help in professional school is novelist and medical student Blake Charlton. "I've discovered that while I'm not a fast reader, I'm a really fast listener. I can listen quite comfortably to a lecture at three times normal speed, and I don't have any trouble comprehending. I can back up when I don't understand the concepts, but I rarely misunderstand the words. With the technology that's out there, I think that learning to rely more on listening when you're young could be a great boon for dyslexics. Recorded books helped me out a great deal when I was younger, but if I could have listened at double speed, that would have been life changing."

The use of technology is sometimes met with skepticism—or even active opposition—by educators and reading specialists who believe that it interferes with efforts to teach children to read. Ben Foss finds this attitude seriously misguided. "This overemphasis on early reading intervention and lack of emphasis on other aspects of reading is just so silly. Using both remediation and accommodation is a much better approach. Let me be clear about this. I believe strongly that there's an appropriate role for phonics training and all that, but sometimes you hit a point of diminishing returns where an hour spent doing therapies isn't nearly as valuable as an hour spent focusing on accommodations, and where the accommodations will do much more to affect long-term outcome.

"I've also found that exposure to language by itself improves literacy, which is your ability to comprehend and access literate information, as opposed to simply knowing how to read. Exposure to language improves your vocabulary, your knowledge base, and your ability to seek content that you find interesting. I'll give you a great example. One kid got an Intel Reader, and he told us, 'I used the Reader to read the rule book for the game Risk, and I found out my friends were cheating!' That kid is now fired up about getting inside of texts because he's always wanted to invade France [in Risk mode only!], and he now knows how. He's discovered that literacy—or access to literature—is power. That's why all kids should start out with both visual and auditory exposure to literature and literate information in all its forms."

In recent years a growing number of prominent educators have reached a similar conclusion. Dr. Lynda Katz is the president of Landmark College, an accredited junior college in Putney, Vermont, with a remarkable record of helping students with dyslexia, attention deficits, and other learning challenges successfully transition into college or professional school. Dr. Katz speaks enthusiastically about the hundreds of students she has observed firsthand benefitting from the use of assistive technologies. "I have an incredible bias for assistive technology. At Landmark, we began using assistive technologies as accommodations, but I really think they're remedial—that is, that their use improves reading function. I've had so many students who've used them who when they get ready to leave Landmark feel like they just don't need them as much. The assistive devices and the exposure to texts they bring really seem to enhance students' literacy skills and improve their reading. Drill and drill and drill just doesn't seem the way to go."

We wish every school would take this more balanced and flexible view of assistive technologies like recorded books and text-to-speech software—particularly for students who are still working to improve their decoding and reading fluency skills. This two-pronged approach gives students with dyslexia the same opportunities to expand their general knowledge base, enrich their vocabulary, and improve their other high-level literacy skills that other students already enjoy. It would likely also—as Ben Foss and Lynda Katz have noted—improve the speed with which they master reading skills.

Finally, parents can play a special role in creating a home environment that encourages greater literacy for struggling readers. Studies have shown that children whose parents engage them in conversations about important and challenging topics at home adapt more quickly to learning from literate texts. Parents should also make sure struggling readers are given access to the material in books by reading aloud to them, and by providing exposure to technologies like books on tape, text-to-speech reading programs, or worthwhile documentary films.

Summary of Key Points on Reading

- Skilled reading requires strengths in decoding, fluency, and comprehension.
- Dyslexic students require extra practice to build decoding skills, and the most effective practice involves explicit training in phonics and phonological awareness.
- Orton-Gillingham–based methods are the gold standard for such practice, though individuals with dyslexia who have more severe difficulties with sound discrimination may also benefit from computer-based auditory training.
- Orton-Gillingham–based methods come in a wide variety of forms, but all achieve their success in part by turning learning into a memorable experience. The particular "flavor" should be chosen to match well with a student's cognitive strengths (including MIND strengths) and interests.
- Fluency training employs both oral and silent reading practice to improve word identification, problem solving, and reading speed. Reading materials should be chosen that grab the student's attention (often by covering an area of special student interest), use straightforward sentences, and employ familiar vocabulary. Dyslexic students' big-picture reasoning skills should also be tapped by providing them context for what they'll be reading (by first reading the passage to them or giving them a brief summary).

- Dyslexic students often possess cognitive strengths that will enable them to become good interpreters of texts, once barriers to accessing the words in the text are removed.
- In addition to reading instruction, dyslexic students should be given access to recorded books and text-to-speech technology, so their exposure to literate information and their cognitive development can proceed at full speed.
- Newer technologies often permit "speed listening," which can greatly improve the usefulness and attention-holding ability of recorded texts.
- Parents can play an important role in turning the home into an environment that encourages literacy for struggling readers by engaging children in challenging conversations and providing access to literacy-building alternatives to print, such as books on tape, text-to-speech reading programs, or worthwhile documentary films.

CHAPTER 26

Writing

Many individuals with dyslexia have the potential to become not only competent but even highly skilled writers. In previous chapters, we've introduced you to several dyslexic individuals who, despite early challenges with reading and writing, have gone on to become extremely talented writers. This pattern is far more common than many people realize.

Typically, when individuals with dyslexia fully develop their writing skills, their mature writing reflects many of the MIND strengths we've described, including the abilities to see distant and unusual connections and associations; view things from different perspectives; see big-picture gist and context; show strengths in scene-based memory and imagery; think in cases and episodes rather than in abstract definitions or generalizations; and engage in mental simulation, prediction, and insight, to see patterns that others often miss. These abilities often begin to show up in the writing of individuals with dyslexia during adolescence and young adulthood, though many dyslexic individuals hit their full stride as writers only during their early to mid twenties, or even later.

Even individuals with dyslexia who ultimately become highly skilled writers will nearly always struggle early in school with the fine-detail features of writing. These difficulties can affect functions like the physical and mechanical skills needed for legible and accurate handwriting; the rule- and sound-based patterns underlying spelling and grammar; the structural and organizational patterns underlying sentence, paragraph, and essay construction; and understanding which details to include and exclude in their writing. For most individuals with dyslexia, fluent and automatic mastery of these skills requires much more time to achieve than it does for other students. Often, it also requires more explicit instruction and greater practice imitating good writing as well.

In this chapter, we'll discuss steps that can help dyslexic students become competent or even highly skilled writers. We'll focus on several important levels of writing and on the role that both technology and the proper use of dyslexic advantages can play in helping dyslexic students develop their writing abilities.

Writing by Hand

Learning to write by hand is often a major challenge for children with dyslexia. Severe problems with handwriting are common among younger dyslexic students. Though Blake Charlton is now an accomplished novelist, he struggled greatly with handwriting: "In special ed class I failed a lot of papers because the teacher couldn't read anything I wrote."

Problems writing by hand can affect any student with dyslexia, but they're often most severe for students with significant working memory or procedural learning challenges. Writing by hand depends almost entirely on automatic, fine-detail skills. These skills allow us to form letters neatly, consistently, and in the proper spatial orientation; properly space letters and words; use margins; and master conventions (like capital letters) and punctuation (like commas and periods). Students who haven't fully mastered these automatic handwriting skills must use conscious attention (working memory) to perform them. As a result, they have less "mental desk space" left over for formulating sentences, organizing thoughts, or checking for errors, so their work is often littered with *overload* mistakes—that is, with more frequent or severe mistakes in spelling, word omissions, inaccurate word substitutions, poor mechanics, improper grammar or syntax, or just general messiness.

Students who struggle to write by hand usually need a combination of specific training (remediation) and accommodations. Training should begin using a program of explicit, rule-based, multisensory approaches to letter formation. Our favorite program is Handwriting Without Tears (www.hwtears.com), which resembles the Orton-Gillingham approaches we discussed in the last chapter by taking advantage of spatial and kinesthetic imagery strengths and using multisensory practice to turn instruction into a more memorable experience. This approach is often administered in schools (sometimes through a school-based occupational therapist), but because it's relatively easy to understand, parents can use it with their children at home as well.

Children with especially poor finger coordination will sometimes benefit from working with an occupational therapist who has been trained to help children with handwriting problems. Many dyslexic children—especially those with procedural learning challenges—show poor fine-motor coordination problems and low muscle tone in the core muscles of their spine, hips, and shoulders, all of which makes prolonged seatwork difficult. These children often benefit from core muscle strengthening in addition to fine-motor (finger) training.

As we discussed in chapter 7, children who invert their written symbols so frequently or elaborately that it holds back their progress in writing or reading (particularly past age eight or nine) deserve special attention. Interventions should be based on the child's MIND strength profile. Since many of these students have tremendous M-strengths (or spatial reasoning and imagery abilities), these talents should be used to minimize symbol flipping. It's often helpful for these students to practice forming three-dimensional clay models of letters and short words (especially words containing the letters that are often flipped). They can also practice letter formation by making very large letters (two or more feet in height) with a marker on a whiteboard or by using their hand to trace out the letters in a box filled with sand or rice. These practices engage their large motor muscle (kinesthetic) memory and activate broader areas of the cortex, which seems to improve spatial orientation. Helpful techniques of this kind are also described in the books *Unicorns Are Real*[1] and *The Gift of Dyslexia*.[2]

Many dyslexic students appear to have preset limits in how neatly and automatically they can learn to write by hand—even with extensive training—and they eventually reach a point of diminishing returns where further progress is hard to achieve. Students who reach this point should be treated with understanding and respect, since their limitations reflect brain biology rather than effort. Even those who do eventually learn to write well by hand often don't achieve full comfort or fluency with handwriting until mid to late adolescence—or even young adulthood. This again is a result of their late blooming in working memory and language, and it's important that parents and educators understand and make allowances for this time frame of development. Fortunately, because many excellent technologies now exist to provide alternatives to writing by hand, there is simply no need for students with especially persistent handwriting difficulties to suffer. We'll discuss these alternatives later in this chapter.

Writing Sentences

Many individuals with dyslexia also have difficulty learning to write at the sentence level. One common problem is learning to master the rules of grammar and syntax, which dictate how words relate to each other and function in sentences. These include the rules for:

- subject and object relationships (*The man walked his dog down the street* versus *The man down the street walked his dog* versus *The man's dog walked down the street*)
- active and passive constructions (*The king kissed the queen* versus *The queen was kissed by the king*)
- tenses (*Yesterday we went to the drugstore, but today we're going to the mall, and tomorrow we'll go to the restaurant*)
- pronouns (*He gave his wife's sister her husband's letter*)
- relative clauses (*The man that is pulling the woman pulls the dog* versus *The man is pulling the dog, and he is also pulling the woman*)

- other grammatical features like prepositions, adjectives and adverbs, multiple word meanings, and complex constructions

As many as half of all college students with dyslexia have been shown to struggle with grammar and syntax—and that only includes students who've done well enough to make it to college! Typically, these students can write sentences with simple "active subject/passive object" formats but struggle as sentences grow more complex.

Experts on dyslexia often don't classify these language issues as part of "dyslexia" per se but as "dyslexia-related language learning differences." However, since these challenges really do appear to result from the same neurological variations as the dyslexic reading challenges we've already discussed, and since mild challenges of this sort occur in many individuals with dyslexia, we believe it's important to discuss them.

These subtle language challenges are often missed on routine evaluations of language function in younger primary school students, so a "clean bill of health" on a brief language evaluation doesn't rule them out. Also, these challenges are often much more apparent with writing than speaking, since writing demands both greater precision and a greater use of working memory resources.

Challenges with working memory and procedural learning are often the chief contributors to dyslexic problems with sentence construction. One of the leading experts on dyslexia-related language and writing challenges is Dr. Charles Haynes, professor of communication sciences at the Massachusetts General Hospital Institute of Health Professions in Boston. When we asked Dr. Haynes where he believed dyslexic students face their greatest challenges with writing, he didn't hesitate. "The sentence is really an overlooked area. When teaching sentences, people think primarily of sentence diagramming, but that's really not what many of our children with dyslexia and related language learning disabilities need. These children often haven't grasped *the sense of a sentence*—or the logic that underlies sentences—so they need practice comprehending and formulating sentences, rather than just diagramming."

Notice how this focus on the logic or meaning of the sentence *as a whole* is a top-down approach. The student begins at the "top," by deciding what purpose the sentence as a whole should perform (e.g., to explain, to describe) rather than at the "bottom," by focusing on details like nouns, verbs, and adjectives. Based on what we've discussed about the thinking styles of students with dyslexia, this top-down or big-picture approach is precisely the kind we would predict would be most effective to help them learn. Yet this approach is almost exactly the opposite of the bottom-up approach that's usually used to teach sentence construction (e.g., start with the noun you want the sentence to be about, now add a verb to show what it's doing, now add another noun to show what it's doing it to, and so on).

According to Dr. Haynes, one key type of practice that students with dyslexia need is to learn the specific sentence formats that are logically linked to particular types of paragraphs. As he told us, "Sentences, especially complex sentences, have a logic. For example, if you form a compound sentence using the word *and*, there's a logic to that: sentences with *and* mean you're talking about *similar events* or *events that follow in sequence.* With all the types of sentences, there are particular sets of words that you use for certain purposes."

These word sets create different sentence patterns, such as sentences about *processes*, sentences that list *reasons* or *traits*, sentences that *urge* or *persuade*, sentences that *describe*, and sentences that *compare* and *contrast*. Understanding the logic of these sentence-level patterns is critical for students when they attempt to combine sentences into paragraphs. "If you want to construct a persuasive paragraph, you need to understand how to form a *because* sentence that expresses cause and effect; and if you're going to write a compare-and-contrast paragraph, you need to be able to create sentences with *although* and *while* because they express that comparative logic."

Before students are asked to write paragraphs, they should master these logical relationships at the sentence level. "If they're still laboring at the sentence level, they're not going to have the cognitive resources available to work at the paragraph level. Paragraphs are really just words and sentences put together in a kind of logical order, and each paragraph type has a kind of sentence type at its core. If students are breaking down at the paragraph level, then we need to make sure they can produce the necessary core sentences. One of our big problems currently is that most people want to hop right to the paragraph level. They say, 'Our state-mandated tests require children to write a personal sequence narrative or a three- or a five-paragraph expository text structure.' So they give kids a paragraph template, but they skip all the important sentence work that leads up to it."

Students with dyslexia who struggle with the logic of sentence construction should begin their instruction in this top-down approach with *oral* rather than written practice. First, they should practice identifying the different logical patterns in sentences (e.g., persuading, listing, describing, comparing and contrasting) as they *listen* to them. Using sentences that relate to areas of student interest or strength can improve focus and endurance. Next, students should be asked to practice *speaking* sentences of these different types (e.g., "Describe this . . ." or "Tell me why I should . . ." or "Tell me three things you want for your birthday . . ."). Next, students should practice identifying these sentence patterns when they *read* them.

Only after students have mastered all these preliminary practice steps should they be asked to actually *write* the different sentence types. Again, the point of this practice is not to learn which words are nouns or verbs but which kinds of words perform various logical functions. We list specific resources for practicing these skills in Appendix A.

Writing Paragraphs, Essays, and Reports

Getting Started at the Paragraph Level. Once students have learned to identify and create different sentence types, they must then learn to construct each of the different types of paragraphs. Mastering the formation of the various paragraph types mostly involves learning to connect sentences of each core type with appropriate introductory and concluding sentences. When students have mastered these steps, they'll have acquired many of the key skills they need to write a full essay or report.

While teaching paragraph and essay construction, it's important to use clear templates. These templates should include both *explicit descriptions of individual steps* and *engaging examples of finished work*. The same *listen—*

speak—read—write progression described above should also be used for paragraph construction.

Again, most students with dyslexia are top-down learners who work best when they know precisely what they're aiming at. In this sense, many students with dyslexia should be thought of as "apprentice" learners who learn best by imitating skilled work: that is, they master their craft more quickly and efficiently when the steps for skilled performance are explicitly stated and demonstrated than when they have to puzzle out the rules for performance themselves.

With paragraph construction, a good example of the paragraph type they're trying to construct should be kept available for reference while they're working. One useful source of templates and instructions is Diana Hanbury King's *Writing Skills* series,[3] which provides excellent descriptions of the various paragraph types as well as a program of graded instruction that will help dyslexic students develop all the important writing skills. Another useful program is *Step Up to Writing* (www.stepuptowriting.com), which teaches students explicit rules about the types of information they should include at different points in a paragraph or essay. *Step Up to Writing* also provides lists of common transition words that dyslexic students can draw upon to link their sentences into paragraphs and essays.

During the writing process, students with dyslexia will often struggle to initiate a new sentence or paragraph, expand their ideas, or find the right words to express their thoughts. When this happens, they should be assisted using a combination of brainstorming, preparation, and prompting techniques. In using these techniques, the student's interests and cognitive strengths should be employed, just as with reading.

Problems initiating writing assignments are often greatest with open-ended questions or assignments. A student's creativity can often be released by negative or contrary prompts—that is, by giving the student a statement or thesis with which he or she is likely to disagree. Humorous or silly prompts often also work well to get the student's ideas flowing. Knowing and focusing on a student's interests and strengths can be helpful in this process.

Students who are particularly strong nonverbal thinkers or visual imagers often benefit from brainstorming that uses sketches, doodles, diagrams, or graphic organization techniques. These can be done on paper, a whiteboard, or a computer. Instruction and practice in formal "mind-mapping" techniques can help individuals use strategies of this type in a more organized and productive fashion. One of the top mind-mapping software programs, Inspiration, was created by Mona Westhaver, who is herself dyslexic and designed this program to meet her own needs: "Visual learning strategies . . . allowed me to capture all my ideas in random order, meaning I was able to exercise my multivariate need to jump around and let lots of information simultaneously flow through my mind and on to the paper. Then, with all of my ideas visually in front of me, I could organize my thoughts."[4] Many individuals with dyslexia experience similar benefits from programs like Inspiration or Kidspiration, which is a graphic organization program designed for younger students. Mind-mapping techniques are also taught in books like *Mapping Inner Space* by Nancy Margulies.[5]

Highly interconnected and insight-based problem-solvers often benefit most when brainstorming is unstructured, relatively free of time pressure, and conducted in a relaxing environment. Very open-ended prompts like "What does this make you think of?" or "What pops into your head when I say . . . ?" often work well. Having a digital recorder or scribe to record ideas without any need for the student to write them down may also help students keep track of their ideas without derailing the creative process. Highly narrative thinkers often benefit from beginning the brainstorming process by searching for cases, examples, fables, legends, myths, or stories that particular questions or topics make them think of. Students should be allowed wide latitude to think freely during this stage of the process.

Other methods for freeing up the creativity of the student before beginning the writing project include:

- Reviewing words that might be used to convey the logic of the sentence or paragraph types chosen. For example, for a compare-and-contrast sentence, words like *while*, *although*, *but*, and *however* should be reviewed. Having a list of these words printed up and visible during the writing process can be a great help. The books by Diana Hanbury King and by Jennings and Haynes in the "Resources for Writing" section of Appendix A provide sources for these lists.

- Having a core set of "go-to" words of different types available during writing—like prepositions, adjectives, and adverbs—and reviewing these lists before beginning to write.
- Picking a particular place or scene or event that the child is familiar with (i.e., one that remains strong in episodic memory) or a subject the child is especially interested in, then using that as the "setting" about which the child will write. (It can also be helpful to review some of the nouns and verbs that are associated with that scene.)
- Brainstorming and reviewing word lists that convey or relate to a particular type of mood, tone, location, topic, etc.

When the student bogs down during the writing process, use prompts like those mentioned above. If the student is struggling to find particular words, Dr. Haynes recommends a "graduated series" of *extrinsic* and *intrinsic cues* like the following:

- Extrinsic cues are hints that a teacher or tutor gives when the student is experiencing word-finding difficulties. A graduated cueing system means that the teacher only gives the student enough cueing to retrieve the word, but doesn't give the word away. A graduated series of cues could include:

    —First a picture
    —Then gesture or a "mimed" hint
    —Then a definition
    —Finally, the first sound or letter

- Intrinsic cues are self-initiated ones, and because of this they are potentially even more valuable in the long run. They include strategies like imagining a verb's action or a noun's appearance, function, location or setting, circumstances, or time of appearance.

In general, writing tasks for dyslexic students should also be "chunked" into small, distinct steps or stages. Each step should be clearly explained and demonstrated to them, then tackled one at a time.

Including—and Leaving Out—the Right Details. Many individuals with dyslexia have difficulty learning how much and what kind of detail to include in their writing. Individuals with strong I-, N-, or D-strengths may include excessive or irrelevant details because they often see so many connections and levels of meaning between ideas. For students who have difficulty narrowing down their ideas, it often helps to decide in advance what the focus of their writing will be. One useful strategy for limiting focus is to use the "5W/H" approach, where the student decides which of the potential questions (i.e., who, what, when, where, why, or how) to answer and which to ignore.

At the other end of the "detail spectrum," dyslexic individuals with either especially strong verbal imagery and/or particular weaknesses in word retrieval or verbal output often include too few details. This can be either because they "see" so much detail in their heads that they forget how little they've communicated to their audience or because it takes so much effort for them to put their thoughts into words that they experience working memory overload before they can get everything down on paper. Students with problems of this kind often benefit from reading their work aloud or being asked to form a mental picture of their subject using only the words on the page.

Reading written work aloud also has other benefits. Law professor and dyslexic David Schoenbrod—who has written four highly regarded books on environmental law and legislation—shared a story that illustrates these benefits: "I had a lot of trouble learning to write. In high school I was assigned to write one thousand words every weekend, and I had a lot of trouble getting finished. My father was an excellent writer, and he was left shaking his head. I labored at writing until a law school professor had us students write short articles about the law in the manner of, say, *The New Yorker*, and then read them aloud to the seminar. That way, we heard our clunkers and might eventually acquire the skill of 'hearing' them even when reading silently to ourselves. This made writing an extension of talking rather than reading. What a relief."

One potential limitation to this "self-proofing" method is that the dyslexic writer's brain often initially "sees" what it thinks it's written rather than what's really on the page. Using a pen or a finger to "tick off" each word read as it is read, or getting the assistance of someone else who can read the work aloud, can eliminate this problem. So can the "read-aloud" (text-to-speech) technologies we discussed in the last chapter.

We sometimes encounter one other rather surprising source of diminished detail in some students with dyslexia. That source is a misunderstanding about the point of school essays. Several dyslexic students have told us they write short essays because they don't like to tell teachers things that the teacher already knows. So they avoid restating facts they learned in school and include only information the class hasn't discussed. These students need to understand that the point of school essay writing isn't to teach the teacher but to write as if trying to teach another student who knows nothing about the topic.

Before moving on we want to stress once more that there's absolutely no reason to push students with dyslexia to write at the paragraph and essay level too quickly. If a student with dyslexia is still mastering sentence-level or single-paragraph writing in middle school, that's okay. Keep working at that level before trying longer papers. Many teachers and parents worry that students will "miss the boat" if they don't develop essay-level writing skills early on. However, the danger is much greater that dyslexic students will give up on writing altogether before they have a chance to "grow into" their talents. Many of the individuals we interviewed for this book were considered poor writers by their teachers in college—including hugely successful novelists Anne Rice and Vince Flynn—yet they now make their living preparing highly polished prose.

The Point of Writing: Using Technology to Put Thoughts on Paper

Just as with reading, educators often disagree about the use of technological accommodations for students who are struggling to write. Some believe that all schoolwork must be handwritten. Others allow keyboarding for certain students but deny other accommodations like oral dictation or oral testing. In addressing these hotly contested issues, our choices may be clarified if we consider the *point* of writing.

Writing allows us to share our thoughts with others in a form that's easy to store and transmit. Writing sharpens our thinking by allowing us to elaborate, explain, and develop our ideas to a greater degree than most of us can do orally. Writing also imposes a greater demand for precision and clarity because it prevents us from "slurring over" our structural and conceptual difficulties as we do when speaking by using facial expressions, gestures, or postures to complete the meaning of our sentence fragments.

These considerations point to several important benefits of writing. The most important of these benefits are the chances writing provides to develop and refine ideas, and its demands for precision and clarity.

Writing by hand is clearly not essential for receiving any of these benefits. So long as your words can be accurately stored and shared, the physical process by which you store those words should be relatively unimportant.[6] In fact, for many students with dyslexia, handwriting is more an impediment than an aid to written communication. When writers haven't yet mastered the automatic formation of letters, spelling, or other skills, handwriting takes up valuable working memory resources that could be better used for more important aspects of writing, like forming thoughts or using proper grammar and syntax. Insisting that students who struggle with handwriting should do all their work by hand is simply unreasonable, unproductive, and when carried to extremes, deeply unkind.

This doesn't mean that students who struggle with handwriting shouldn't practice to improve their handwriting. On the contrary, dyslexic students who struggle with handwriting should practice every day using the techniques we've described, so their handwriting will eventually become more automatic. *But students whose handwriting is not yet automatic should learn handwriting and written communication as two separate subjects.* Handwriting practice involves forming letters, writing words, and using written conventions. Writing practice involves constructing sentences, paragraphs, and discourses (essay, story), and for students with severe handwriting difficulties this is usually best done using keyboarding (if possible) or dictation.

Keyboarding has many benefits for dyslexic students besides making their writing easier to read, so it can actually be valuable for all individuals with dyslexia—whether or not they have difficulty writing by hand. For students who are even modestly proficient with typing, keyboarding requires less working memory than handwriting. As a result, more working memory resources are left available for other aspects of writing. Cut-and-paste functions also make editing and rewriting easier, which is important since individuals with dyslexia almost always need to proof and polish their work. By lowering the effort needed to revise, word processing makes it dramatically easier for individuals with dyslexia to produce documents they can be proud of and that will earn higher marks from their instructors.

Word processing programs with an interactive spell-check function—especially programs that are geared for dyslexic students—are also very helpful. As Blake Charlton recalled of his middle school years, "I was given a calculator and spell-check for everything, and almost overnight I went from just barely passing my exams to being very far up on the curve."

Not only do these programs help dyslexic students reduce spelling errors, but when used consistently they actually teach individuals with dyslexia to spell better. Spell-checkers provide immediate feedback on spelling mistakes—which is most valuable for producing lasting change and learning. They also focus their feedback on words that students actually use. We've often seen remarkable improvements in the spelling of dyslexic students who regularly use spell-check—even if they've made little progress with more direct and traditional forms of spelling instruction.

Word processing programs that are especially helpful for students with dyslexia also provide functions like grammar checking, word prediction functions that "guess" the word the student is trying to spell, read-aloud dictionaries that pronounce highlighted words and state their meanings, and oral text-to-speech (or read-back) functions that "tell" students what they've written. While even standard word processing programs like Microsoft Word come equipped with some of these functions, other programs have been designed to be particularly useful—and user-friendly—for individuals with dyslexia. We've listed some of the best of these programs in Appendix A.

There are two more kinds of technologies that many individuals with dyslexia find helpful for important writing needs. The first is speech-to-text software, which allows the writer to orally dictate into a computer microphone and then the program translates his or her speech into printed text. We've found these programs to be very helpful for individuals beyond mid adolescence, although younger students typically have difficulty making them work well and generally do better dictating to a parent, tutor, or other scribe.

The second form of technology helps with note taking—a task that's often especially challenging for students with dyslexia. While recorders have long been the standard option for students who struggle to take notes, one more recent and especially creative system combines a special paper notebook and an mp3 recorder to create an information capture system that allows students to take highly functional notes with only a minimum of writing. Many of our students have been quite successful using this product, which its manufacturer, Livescribe, calls a smartpen. See Appendix A for more on this and the speech-to-text tools mentioned above.

Ultimately, these accommodations are helpful only if students agree to use them, and unfortunately many students with dyslexia reject accommodations for fear of looking different or of being accused by classmates of cheating. Brown University graduate and dyslexia advocate David Flink described to us his discomfort as a young student using a laptop in class. "As a kid with dyslexia, I was different always, and that laptop was just one more mark on me after so many others that said I was broken and I couldn't do things the way everyone else did. It made me feel so much less resilient. Even though I felt empowered with my laptop, I also felt like I was somehow cheating and breaking the rules. In the end, though, I decided I was still willing to 'break the rules' and use the laptop, because it was better to be a smart kid with a laptop than a dumb kid without one. I always knew that I could lay down my laptop and pass as not dyslexic, but then I'd fail."

When students with dyslexia resist using accommodations, it's important to speak frankly with them about the fact that they really do have needs that are different from many of their classmates'—and *that's okay.* Accommodations aren't cheating, and they don't unfairly plant ideas into anyone's head. They simply remove the barriers that prevent dyslexic students from expressing what they already know—and that's all. Appropriate accommodations are often essential for students with dyslexia to develop their skills as writers, and that unlocking of potential is ultimately what education should be all about.

Summary of Key Points on Writing

- Learning to write is especially difficult for many students with dyslexia because they will often struggle with handwriting, expressing thoughts in words, or combining words into sentences and paragraphs.
- With respect to handwriting, students with dyslexia often require more prolonged and explicit instruction in skills like letter formation, spacing, and use of conventions. Multisensory programs that engage imagery strengths and turn learning into a memorable experience are often the most beneficial in helping students gain automaticity in these areas.
- For students who lack automaticity in handwriting, *handwriting* and *written expression* should be treated as separate subjects and practiced independently. In other words, handwriting should be practiced *as handwriting*, while ideas should be communicated by oral dictation or keyboarding (for students who are fluent in this skill).
- Students with dyslexia often struggle with sentence formation due to difficulties mastering syntax or the logic of different sentence types.

Explicit instruction in these skills is often required, and students should achieve sentence-level mastery before moving on to paragraph- or essay-level writing.

- As with reading, the interests and special abilities of students with dyslexia should be considered and engaged in all assignments.
- Keyboarding is extremely useful for all students with dyslexia, and they should use it to write all passages above the sentence level. Not only does keyboarding lower the working memory burden imposed by writing for the many students with dyslexia, but it also provides helpful functions like read-aloud (which can be used to proof work), cut-and-paste, and spell-check, each of which greatly lowers the burdens of revising and polishing work. Keyboarding using software equipped with spelling and grammar checking also has valuable educational effects because it provides immediate feedback on errors for struggling students.
- Dyslexic students should be helped to recognize that they have needs that are different from many other students', and they should accept appropriate accommodations.

CHAPTER 27

Getting a Good Start: Elementary through Middle School

Children with dyslexia face two special challenges during the years from birth through mid adolescence: mastering the basic brain functions that underlie reading, writing, and other academic skills; and developing a healthy self-concept and the strong and resilient character that comes from it. Meeting both these challenges requires careful balancing because each creates demands that often come in conflict.

For example, no one would deny that children with dyslexia should begin intensive training as soon as their problems with reading, spelling, and writing are detected, both because training in the first decade of life is generally more effective and because early intervention can prevent years of academic and emotional difficulties, like poor grades, loss of self-confidence, underachievement, misbehavior, and depression. Yet if we focus too heavily on fixing the weaknesses of children with dyslexia, we may fail to foster their strengths. This is one of the biggest problems with our current educational system—and a key reason why so many students with dyslexia emerge from their early school years feeling scarred and defective.

These crucial early years, from kindergarten to mid adolescence, are when the battle to develop confidence, resiliency, and a positive self-image is largely won or lost. If a student with dyslexia can reach the age of fourteen or fifteen with a healthy sense of self-esteem and a realistic acceptance of both personal strengths and weaknesses, that student is much more likely to enjoy a happy and successful life. The question is, How can we help dyslexic students navigate this challenging period with both their intellectual and emotional development on track?

Several answers to this question were suggested by a fascinating twenty-year study of graduates from the Frostig School in Pasadena, California, which specializes in teaching children (grades one through twelve) with learning challenges, including dyslexia.[1] In that study, researchers identified several key factors that distinguished graduates who were thriving (as classified by several factors including personal satisfaction, career success, and relationships) from those who were struggling. These key factors included a realistic self-awareness and acceptance of learning differences; adaptive personal skills like perseverance, proactivity, goal setting, and emotional stability; and an effective support system.

Many of the individuals with dyslexia we interviewed for this book named similar factors when asked what they believed was critical for their emotional and professional success. These factors included tenacity; confidence; positive self-image; a realistic acceptance of the personal struggles and shortcomings associated with dyslexic learning challenges, but also a deliberate focusing on personal strengths and areas of special interest; supportive home and school environments; and a supportive network of friends.

All factors on both lists fall into roughly two categories: *internal supports* and *external supports*. Let's look at each.

Internal Supports

The first internal support is the sense of *confidence and self-worth* that individuals with dyslexia develop when they learn to recognize and use their personal strengths. Each of our successful dyslexic interviewees mentioned how important they'd found it for the development of their confidence to be able to use and develop their talents during their years of academic struggle.

Many also mentioned the importance of having their talents recognized by others.

These talents were of many different kinds, but most were demonstrated far more easily outside the classroom than in it. Many (though not all) reflect the MIND strengths we've discussed. For example, James Russell and Lance Heywood enjoyed positive experiences working with electronics. Lance's son Daniel and James's grandson Christopher enjoyed positive experiences building elaborate projects with LEGOs and robotics. Jack Laws said positive experiences with Scouting and nature studies "helped me have more self-confidence." Douglas Merrill and Anne Rice cited the importance of their strengths in storytelling. Blake Charlton said sports and drama were "the only ways I maintained my self-esteem." Vince Flynn mentioned sports and chess. Ben Foss cited sports and student government. Glenn Bailey mentioned sports and recognition for his sense of humor. Sarah Andrews and David Schoenbrod mentioned awards they'd won for art.

Similar blends of aptitudes and interests are also common among the students with dyslexia with whom we work. Some of the most common interests we find in these students include (in no special order): nature, animals, photography, videography, animation (hand or computer), art, robotics, LEGOs, storytelling and creative writing, debate club and speech competitions, chess club, science club and fairs, collecting, gaming, planes, cars, motorcycles, boats, electronics, physics, music, crafting, shop, engines, landscaping, dance, sports, inventing, design, fashion, skateboarding, snowboarding, theater, martial arts, computers, Scouting, religious youth groups, kite flying, literary interests (mythology, fantasy, science fiction, historical fiction), history (military history is especially common), and building general or specialized knowledge by listening to books on tape.

We've listed these interests and activities to give you some idea of the breadth of the possible areas of interest through which students with dyslexia can experience confidence and success. Depending on the child's interests and aptitudes, the activity pursued may be cooperative or competitive. If competitive, it's important to find a place where the child faces an appropriate level of challenge. Not everyone can count on being a naturally great athlete, artist, or musician, but exceptional ability isn't essential to experience success. Finding a less elite training environment where children can compete with others at a similar level can allow them to gain confidence while acquiring skills and experience. We've seen many children with dyslexia thrive in martial arts courses that stress personal discipline rather than elite competition and where, when sparring is practiced, it is done with carefully matched children, so matches are appropriately competitive.

Canadian entrepreneur Glenn Bailey shared a story that perfectly describes the importance of finding the right environment to build skills. "As a kid I played hockey in West Vancouver, and I thought I was terrible. But we moved to Vancouver Island [which is much more sparsely populated] the year they opened a new hockey arena, and I became one of the best players on one of the teams because it was a 'new market' and no one else could even skate. So I played on an all-star team, and we traveled all over the island, and I developed all this confidence. And some years later I ended up going back and playing in West Vancouver, and by that time I was one of the best players. That's just a confidence issue, but that's so applicable for a dyslexic person because their underlying issue is often a lack of confidence, which results from the failure they've encountered in the classroom—and confidence is everything in life."

Dr. Charles Haynes, whom we met in the last chapter, has worked with many students and adults with dyslexia as a language specialist, researcher, and classroom teacher. He summed up the importance of focusing on the child's strengths in a way that wonderfully echoes Glenn's observations. "The child with dyslexia has strengths that need to be recognized and supported early—just as early as they need help with their areas of need. They need early positive experiences with people who have faith in them: people who believe they have something to offer and who don't just focus on their areas of need but also celebrate and publicize appropriately what they've accomplished—not in an unrealistic or artificial way but appropriately and sincerely. When children with dyslexia experience success and recognition, they have the confidence to deal with their difficulties in school." This connection between success, confidence, and motivation is absolutely critical to understand for those trying to teach and raise children with dyslexia.

The second internal support is an *attitude* characterized by *optimism and a strong belief in a bright future*. Students with dyslexia are constantly at risk for being overwhelmed by reminders of their apparent inferiority. Each day they're surrounded by classmates who are acquiring skills more quickly and efficiently, scoring higher on tests and assignments, producing neater and longer papers, and finishing tests and assignments more quickly. Students with dyslexia not only feel that they're being left behind, but they see the distance separating them from their classmates increasing at an accelerating pace. Faced with this constant barrage of negative messages, students with dyslexia are constantly at risk of losing their hope for the future and entering a self-defeating cycle of pessimism, loss of motivation, underachievement, and behavioral and emotional dysfunction.

Psychologist Martin Seligman has described how a pessimistic mind-set develops, and the problems to which it can lead, in several classic books like *The Optimistic Child* and *Authentic Happiness*.[2] According to Seligman, individuals who encounter repeated failures often begin to experience a sense of powerlessness. This leads them to attribute their problems to factors that are permanent (or unchangeable), pervasive (affecting not only the areas where the failures occurred but every aspect of life), and personal (or due to some defect within themselves, which they believe to be inescapable or even deserving of punishment). This *pessimistic interpretive framework* then becomes a self-fulfilling prophecy as demoralization (or even clinical depression) reduces motivation and effort, which leads to further failure and an apparent validation of the pessimism.

Fortunately, Seligman has also shown that an optimistic interpretive framework can be taught and learned, so it can replace the pessimistic one. The optimistic framework attributes failures to factors that are temporary and changeable rather than permanent, specific to particular tasks rather than pervasive across all areas of a person's life, and attributable to factors that have nothing to do with an individual's personal worth.

We've found that teaching this optimistic framework to students with dyslexia can help them interpret and deal with their dyslexia-related challenges in more productive ways. This involves teaching individuals with dyslexia that their challenges are temporary and conquerable (either through the use of remediation, strategies, or accommodation), limited to particular functions (which are also accompanied by benefits), and due to specific patterns of brain organization and function rather than to a lack of effort or merit on their part. When they fully grasp the truth of these messages, the results can be transformative.

Individuals who've been deeply discouraged by past experiences may benefit from working with a clinical psychologist or therapist who's been trained in cognitive behavioral approaches. These approaches work by teaching students to change their interpretive framework and to deal more productively with their experiences.

Students with dyslexia should also be encouraged to use the present moment to practice and prepare for the future, rather than to wallow in the defeats of the past. They must avoid at all cost dwelling on past failures—which they can do nothing to change—and should instead focus on building skills that will allow them to avoid similar mistakes in the future.

Another way to encourage students with dyslexia to focus on the future is to have them regularly prepare lists of realistic and achievable future goals—both in areas of challenge and in areas of strength. Goal setting was cited as an important factor in the Frostig School study and by many of our interview subjects. We've mentioned, for example, how Jack Laws's goal of creating a "perfect field guide" helped motivate and sustain him through years of difficult study and how Glenn Bailey credited success in business largely to his practice of long-term goal setting.

A third internal support that can also help students with dyslexia develop a more optimistic outlook is training in *metacognition*, or "thinking about thinking." In particular, they can benefit from learning to understand what's different—and desirable—about dyslexic minds. Students with dyslexia can be helped to adopt a more optimistic and resilient perspective if they're taught what we've described in this book about the different thinking and processing styles that individuals with dyslexia exhibit, their tendency toward late-blooming development, and the life experiences of talented individuals with dyslexia. Understanding—and accepting—their special needs and abilities will help students with dyslexia view themselves in a more positive light, and it will also help them advocate for themselves in a more positive and productive way.

Glenn Bailey beautifully summed up the benefits he experienced when he decided to learn more about dyslexia. "I learned early on that since you can't get out of dyslexia, you'd better get into it. So I decided to 'embrace the beast' and study dyslexia. I learned what it was all about, and how to get the most out of it, and how to be proud of who I am. I learned that there really are a lot of amazing people who are dyslexic and all about this amazing creativity that seems to be a part of it. One thing I really want to do is help people who are dyslexic—including some of my own kids—realize that they have these great things to offer. That's my gift back to society."

External Supports

For children with dyslexia, the support they receive from parents, teachers, and the right school environment is also critical in getting off to a good start.

Many of our interviewees spoke gratefully of the support they'd received from parents during their critical formative years when their self-concept was most vulnerable. Blake Charlton spoke of how important it was for him that his parents recognized and praised his hard work, even when he was earning poor marks in school. Entrepreneur Douglas Merrill recalled the endless hours of tutoring in math that his mother gave him all the way through high school. Civil rights advocate Ben Foss stressed the importance of his parents' decision to pay more attention to what he could do than what he couldn't. "From a very early age my parents looked for ways to find my strengths and encourage them, and they started me out in life with the assumption that you don't have to do things the way everyone else does. I think that freed me tremendously to experiment with different ways of approaching knowledge." Ben's mother also continued to help him proof his papers all the way through law school. He described to us how he would fax his papers home and his mother would read them aloud to him over the phone to help him identify his mistakes. Speech professor Duane Smith also praised his parents for helping him maintain his optimism for the future. "One thing I always knew was that my parents loved me unconditionally. To this day they're still my biggest fans. They always knew that somehow, some way, I would accomplish something. They didn't know how it was going to happen, but they always made me feel that they knew it would someday, and that has helped make me who I am."

Teachers also played a critical role in building the self-esteem of many of our interviewees. We mentioned earlier how Jack Laws credited two of his high school teachers with turning him around as a student by recognizing the quality of the thinking that lay behind the superficial errors in his writing.

Duane Smith also spoke of the important role played by one very special teacher. Duane was making his fourth attempt at community college when he finally met a teacher who was able to spot his special abilities. "We were in a classroom and I delivered a speech for her, and she looked at me and said, 'You have such presence.' That was the first praise I had ever been given in my twenty-one years of attending school, and it was empowering. That phrase became my internal mantra: if I felt discouraged or overworked or tired or had sudden stage fright when I had to give a speech, I'd just repeat, 'Betty says you have presence. Betty says you have presence. . . .' If not for that teacher, I would never have joined the speech team, and I probably would have continued bouncing around L.A. as a bartender or a salesman, or as somebody who was hanging out in nightclubs every night of the week, and who knows where I'd be now." Teachers should never underestimate the power that even a passing word of sincere praise can have to motivate and inspire a student with dyslexia. Often these students are so starved for praise that even the slightest bit of encouragement can do wonders.

Finding a Supportive Environment for Education. It's also essential to find an educational environment that's a good fit for each child with dyslexia. Children with dyslexia differ tremendously in the kinds of environments they find intellectually and emotionally nourishing, so there's really no single educational "size" that fits all students with dyslexia equally well. Instead, it's a matter of finding "the best school" or "the best teacher" for each child.

Usually, the student's own response to a particular educational setting is the best guide to the quality of its fit. In good-fitting environments, children with dyslexia are challenged, but the challenges are matched to their individual needs, abilities, and states of development, and they're increased one small step at a time so that goals remain obtainable. When children with dyslexia are challenged in this careful fashion, the result is typically positive. After a brief but essentially inevitable period of frustration and discouragement, the child begins to make progress, and this progress builds confidence that sustains the child through further slowly increasing challenges.

In contrast, when children with dyslexia are given challenges they cannot meet, their frustration persists and they are at risk for reactions like stress and anxiety, anger, misbehavior, demoralization, even clinical depression. If prolonged, these responses can become a lasting part of a child's emotional and behavioral makeup. It's important to remember that the nervous system treats anxiety and depression just like it treats any other "skill": the more you practice them, the "better" at them you become. For example, the more you "practice" stress, the less it takes to make you feel stressed out, and the longer you remain feeling stressed. Stress also has a highly negative impact on learning because it lowers working memory, focus, and motivation. That's why keeping children in environments where they feel chronically stressed is emotionally harmful and educationally counterproductive.

To better understand the effects of the academic challenge we place on children with dyslexia, think about how we train the bodies of young athletes. We help young athletes build their strength by having them start by lifting light weights, then we increase the weights gradually as their progress and development allow. We'd never begin by loading a three-hundred-pound weight on a seven-year-old child and expect that it would produce strength. At best this would cause frustration and failure, and at worst it would cause serious injury. It's surprising that we expect better results when we give children with dyslexia unbearable academic burdens. We place demands on them that they can't possibly meet, then react with astonishment when they become frustrated, anxious, inattentive, bored, depressed, unruly, or overactive. But this response is inevitable, and the fault is ours, not theirs.

As we've shown you throughout this book, children with dyslexia are "programmed" to develop in good ways but along different paths and at different rates from other children. Attempting to bend and prune them to fit the educational programs we've designed for very different kinds of learners is both harmful and unreasonable.

Students with dyslexia need a kind of education that's designed specifically for them, an approach that spends as much time enhancing their strengths as it does diminishing their weaknesses. Students with dyslexia need an educational environment that teaches them about things that interest them in addition to helping them learn to read and write. Unfortunately, in our obsession to have them master basic skills at younger and younger ages, we often fill the days of our young students with dyslexia entirely with the kinds of rote and procedural tasks that they find most difficult. As a result, their days become times of unrelieved frustration and failure, and before long their motivation and self-confidence drain away.

It's essential that we give children with dyslexia learning environments that balance basic skills training with exposure to fascinating information about our world. A few "mainstream" schools do a good job of this, and if students with dyslexia are given the extra help they need to address their challenges along with appropriate accommodations in the classroom, they may do well in such environments. This is especially true of children who are naturally blessed with highly resilient temperaments and a strong sense of self-confidence, as these strengths allow them to cope well with the often highly visible differences that set them apart from their classmates.

However, for children who remain constantly aware of—and discouraged by—their differences from classmates, remaining in a conventional classroom can be emotionally devastating. In our own clinic, a shockingly high percentage of early elementary students who come to us with reading and writing challenges have expressed thoughts of death and suicide. For older students, these numbers remain alarmingly high. These emotionally vulnerable children often do much better in a setting where they're grouped with other children with similar academic challenges. Several of our interviewees spoke of the extra confidence they received from spending several of their early years in a special education classroom where they were viewed as relatively successful.

For some students, attending schools that specialize in teaching students with dyslexia or other learning challenges can be a good alternative. David Flink, who heads a remarkable nonprofit mentorship program for individuals with dyslexia called Project Eye-to-Eye (which we'll discuss in the next chapter), described the benefits he experienced in going to a school devoted to children with learning challenges. "In fifth grade I was finally diagnosed as dyslexic, and I started going to a school specifically for kids with learning disabilities. At that school I was given a ramp into learning, and that experience helped me realize that *I* wasn't what was broken: it was what was being used to teach me that was broken. I never changed, but the way I was taught changed. That realization made all the difference. In the two years at that school I learned to read . . . and it was an amazing ramp into books. I left that school feeling incredibly empowered."

There are many excellent private schools that specialize in teaching children with dyslexia and other learning challenges, and we keep a list of several at our Dyslexic Advantage website. Some general education private schools may also provide a good fit for students with dyslexia if they offer flexible or individualized alternatives for students to work at their own pace.

Finally, we have seen some students with dyslexia flourish with homeschooling. Suitably equipped and motivated parents can sometimes provide all the teaching the child needs, but often the addition of private tutoring in more difficult subjects like phonics and writing can be helpful. Homeschooling offers a number of benefits that can be especially helpful for students with dyslexia. It removes the stress of comparing personal progress with peers, it allows more time for focusing in depth on special interests, and it allows children who are ahead in some subjects but behind in others to pursue more advanced studies as their ability permits. There are now many excellent alternatives for online learning that can be pursued at home as well. More information on homeschooling options is also provided on our website.

In closing, we want to stress again how important it is to pay attention to the response of a child with dyslexia to his or her learning environment.

Never assume that a child who is showing resistance or acting out in response to a particular lesson, curriculum, or classroom is simply shirking. Children crave success, and it's in their nature to learn and grow. If they reject what we are offering them, that rejection is often a form of defense that they're using to avoid failure when they feel that success is impossible. However, when provided with an environment that's appropriately nurturing, and where success is both possible and praised, most children will respond with greater motivation, effort, and interest. That response is essential to helping them get off to a good start during the early years of school.

How to Help Students in Elementary and Middle School

- During their early years, it's every bit as important to make sure that children with dyslexia develop a healthy self-concept as it is to make sure that they develop basic skills in reading and writing.
- A healthy self-concept can be fostered through the right combination of internal and external supports.
- Internal supports include:
    - —The self-confidence that comes from focusing on and developing strengths
    - —A sense of optimism (or optimistic interpretive framework) and a firm belief in a bright future
    - —An understanding of how one's own mind works, including what's unique and special about how dyslexic minds function
- External supports include the care received from parents, teachers, and the right school environment.
- The appropriate educational environment must be determined for each child. All good environments provide challenges that are doable and that increase incrementally as progress allows.
- Remediation in areas of weakness must be balanced by interesting and engaging work in areas of strength if students with dyslexia are not to lose heart.

- The child's own response is often the best indicator of whether the right balance has been achieved and the right environment identified.
- The stress response is treated by the brain like any other "skill": the more it's "practiced," the stronger and more long-lasting it becomes. Children who show signs of significant stress in school must be treated with great care and attention.

CHAPTER 28

Thriving in High School and College

The period from mid adolescence to young adulthood is also a critical time for individuals with dyslexia. During these years they must become increasingly responsible for their own organization, learning, and significant life choices.

One of the most important choices facing individuals with dyslexia during this time is whether to attend college or to head directly into the workforce. In this chapter, we'll focus on the decisions and challenges facing those who choose to go to college. We'll discuss issues related to work in the next chapter.

Developing the Skills and Supports Needed in College

Students with dyslexia who plan to attend college face two important tasks during their high school years. The first is developing the skills and supports they'll need to succeed in college.

Learning and Study Skills. Students with dyslexia must first and foremost develop the ability to identify and use their ideal learning style. An individual's ideal learning style is determined by his or her blend of four key learning components. Those components are *Information Input*, *Information Output*, *Memory* (or *Pattern Processing*), and *Attention*.

Information Input refers to the routes through which we absorb information. Some students take in information best through auditory routes and are good at remembering things they hear, while other students learn almost nothing by listening and find lectures a waste of time. Some students (even some with dyslexia) learn best by reading, while others learn poorly through print. Some students learn well by looking at visual representations of information, while others must put things into words to remember them. Others learn best by interacting physically with information or learning through exploration, while others find that activity distracts them from learning. Each student must find the routes that work best for him or her, then do everything possible to channel incoming information through these routes.

Information Output refers to the routes through which we express or communicate information. Some students are powerful oral communicators and can easily express their thoughts by speaking. Others communicate better by putting their thoughts in writing. Others express their thoughts best using visual or structural representations, like diagrams, schematics, or working models. Finding educational environments where work requirements are well matched to output strengths is also essential.

The third learning component, *Memory*, we discussed in chapter 16, but understanding memory is so important for determining ideal learning styles that we'll touch again on some key points. During this discussion you can refer to figure 1 on page 115, which illustrates the structure of the memory system.

Memory can be divided into two main branches: working and long-term memory. Working memory is like the random-access memory (RAM) on your computer. It's where information in current use is kept so it can be quickly accessed for processing. Working memory has visual, verbal, and spatial/kinesthetic branches, and different students may show big differences in how much information they can hold in each. Knowing which branch of working memory works best for them can help students with dyslexia channel information into the appropriate form. For example, students with strong visual working memory can turn all sorts of information into visual representations like charts, graphs, icons, pictures, or mind maps. Students with strong auditory-verbal working memory may use key words or acronyms to hold a larger amount of information in a smaller working memory space. And students with strong spatial/kinesthetic working memory can use movements or positions in space as "pegs" to keep information in mind. We discuss such memory (or mnemonic) techniques in detail in our book *The Mislabeled Child.*

It's important to realize that students with working memory weaknesses—which includes many individuals with dyslexia—can also "offload" their working memories by using external memory aids or "surrogate memory devices." These aids can include word lists of key terms; cards showing formulas, steps in a type of problem, or examples of the type of problem being solved; lists of sentence and paragraph types (as discussed in chapter 26, on writing); checklists of "to do" items; or any of the other organizational methods and technologies we've been discussing. With the right strategies to offload working memory and take advantage of other cognitive strengths, limitations in working memory don't need to cause serious problems.

The other major branch of memory, long-term memory, can also be divided into two branches: procedural memory and factual (or declarative) memory. As we've discussed, procedural memory helps us automatically master procedures, rules, and rote tasks so we can perform them without consciously thinking about them or using working memory. Inefficiencies in procedural memory are common in individuals with dyslexia, but just as with working memory, problems can often be prevented by using appropriate strategies and accommodations to offload procedural memory. These strategies involve explicitly studying and practicing the rules and procedures for complex tasks and having memory aids containing this information available when they are being practiced.

Factual or declarative memory can also be divided into two major branches. The first branch of factual memory is episodic or personal memory, which we described in detail in chapter 16. This is memory for things as they've been personally experienced, or imagined as experiences. Episodic memories are typically recalled as mental scenes that are reconstructed in the mind using bits of past personal experience.

Semantic or impersonal memory, the second branch of factual memory, contains information in a form that's noncontextual or unrelated to specific experiences. Semantic memories are typically more like abstract definitions than examples.

As we've written earlier, in our experience most—but not all—individuals with dyslexia favor episodic over semantic memory. Knowing which form of factual memory a student prefers can greatly help with learning. For example, students with a strong episodic memory often remember facts better when they couch them in story format, whether the stories are real or fanciful (like Blake Charlton using fanciful narratives to represent the periodic table). Individuals with strong episodic memories also tend to remember information better when they think in terms of cases or examples rather than abstract or noncontextual definitions. In contrast, students who favor semantic memory will do best when they can "boil down" specific examples into general principles or underlying themes.

Attention, the fourth system component, is often spoken of as if it were a single function, but it is actually a complex combination of different subsystems. Many students with dyslexia will struggle with some aspects of attention in certain situations, so it's important for these students to understand how attention works in order to troubleshoot their areas of challenge.

One of the key components of attention is working memory, which we've already discussed. Individuals with relatively small working memory capacities (including many individuals with dyslexia) often experience lapses in attention during tasks that place heavy demands upon working memory. What's actually happening is that they're experiencing a breakdown in attention because their working memory is being overwhelmed. This type of working memory overload and attention breakdown also occurs often in individuals with procedural memory inefficiencies because they must perform many more tasks than most people using conscious focus and working memory, since these tasks have not yet become automatic.

Other key components of the attention system include sustained attention, or the ability to remain focused on a task for long periods of time, and selective attention, which is the ability to focus on one thing and to resist distractions. Attention is also heavily affected by factors like motivation and interest, temperament (especially resistance to frustration), and difficulties with information input and output.

Understanding attention is important because many students with dyslexia show big differences in their ability to focus on information of different kinds or in different formats or in different settings. By optimizing the form and setting of the learning experience, students with dyslexia can often greatly improve their attention and learning.

When all four of these learning components—input, output, memory, and attention—are optimized in this way, the effect on learning can be truly dramatic. We discuss these learning components and the ways in which they can be optimized in detail in *The Mislabeled Child*.[1]

Reading and Writing. Just as no one should go exploring in the wilderness without the necessary supplies, no student with dyslexia should enter college without a fully developed plan for dealing with the inevitable reading and writing requirements. Students with dyslexia should begin to develop this plan in high school, and it should include steps both to build their skills in reading and writing and to familiarize themselves with the accommodations and technologies they'll need to succeed in college. We've discussed reading and writing at length in previous chapters, so we won't repeat all that information here. However, we strongly encourage all college-bound students with dyslexia to carefully consider that material and make sure they develop the skills and identify the technological assists and other accommodations they'll need to succeed in college. The students with dyslexia who thrive in college are proactive and head off problems before they occur. Often there won't be time available to address these challenges in the middle of a busy college term.

Organization and Time Management. Staying organized and using time efficiently are also key components to achieving success in college. Traditional methods of keeping organized and on schedule, like checklists, whiteboards, Day-Timers, and sticky notes, can all be helpful. Even more valuable for many of today's increasingly "wired" and "plugged-in" students are the newer technologies that provide reminders through computers or cell phones, or desktop timers for computers that help improve time awareness and focus during tasks. Examples of these technologies are listed in Appendix A. Suggestions for other devices can also be found at the Lifehacker website (www.lifehacker.com) or on our website (http://dyslexicadvantage.com).

Peer Support. As students mature, their relationships with friends become an increasingly important source of support and self-esteem. Unfortunately, only a few of the individuals with dyslexia we interviewed received support from other dyslexic individuals during high school or college. Most remembered those years as a time of isolation and loneliness when they felt separated from the "normal" students by their academic struggles and from other individuals with dyslexia because they were "mainstreamed" into regular education classrooms and had no way of identifying each other.

However, one of our interviewees found his life so wonderfully transformed by the supportive community of other students with dyslexia whom he met in college that the chief mission of his life has become extending that support to other students with dyslexia. David Flink described his early years of schooling this way: "As a kid I always remember feeling alone. Until I got diagnosed with dyslexia in fifth grade, I just felt dumb. The words on the pages didn't make sense to me. Then in fifth grade I was diagnosed, and I started going to a school specifically for kids with learning disabilities."

While that school taught David how to read and write, after two years he was returned to a general education classroom, where he again felt isolated. That sense of isolation continued until he enrolled as a freshman at Brown University. There his life began to change in a way that he'd never imagined. "Prior to enrolling at Brown, I'd never known anyone else who felt and seemed really smart but who also had a learning disability, because everyone I'd known with learning disabilities had lost so much self-esteem that they didn't feel smart at all."

But at Brown, David found an entire community of smart individuals with challenges like dyslexia and ADHD. "When I showed up at Brown, the first week of school the disability office held a meeting for all students with learning disabilities, and at that meeting I met a phenomenal group of people. We immediately bonded and started hanging out together. We quickly realized that one thing we shared was that we'd all been told in one way or another that college wasn't in the cards for us—yet here we all were. So we created what we jokingly named the 'LD/ADHD Mafia' [for "learning disability"/"attention deficit/hyperactivity disorder"]. It was our own version of a 'secret society,' and we shared each other's gifts and skills and supported each other when it came time to ask for accommodations. That community helped us realize that we weren't broken, but that the system was broken. Embracing my own identity as someone who had a learning difference—and using that identity in a positive way—was huge for me. That was how I first realized that dyslexia could be an advantage."

David and his friends benefited so much from their community that they began to realize that other students with learning challenges could also benefit from similar communities. From this realization, Project Eye-to-Eye was born. This is a nonprofit organization (www.projecteyetoeye.org) that David cofounded with fellow dyslexic and "Mafia" member Jonathan Mooney. (Jonathan is also coauthor of the wonderful book *Learning Outside the Lines*, which has become a classic resource for college-bound students with learning challenges.[2]) It is a mentoring program that pairs college students with learning challenges who've successfully learned *how* to learn with struggling secondary school students, so the older students can come alongside and support with advice and encouragement. As David explained, "We realized that it was important to go back and tell younger students that they can be successful and make it to college, too. When you're in college, there's an innate coolness about you for younger kids: you're only one educational leap above them, and when you actually go to them and say, 'You could be like us,' they can actually reach out and touch you and see that you're real, and that's incredibly powerful for them. So the experience of going and mentoring these kids and giving them a different message than they were currently receiving was incredibly life changing not only for them but also for us."

Project Eye-to-Eye has grown rapidly in the years since David and Jonathan established the program. Currently, there are chapters at forty universities and colleges across the nation, with more being added all the time. Even for schools without a chapter, parents, teachers, administrators, and dyslexic students can learn from Project Eye-to-Eye about the importance of building community among students with dyslexia.

Applying to College: Special Considerations for Students with Dyslexia

When students with dyslexia decide to attend college, they should approach this transition with careful planning and strategy. Some of this planning should start as early as middle school. The following points are important to keep in mind.

Choosing the Right College, at the Right Time, for the Right Reasons. Students with dyslexia should choose their college as practically and dispassionately as possible. Each college should be evaluated for the help it can provide in reaching further goals, rather than approached as if getting in and going to that particular school were the goal in itself.

Factors Affecting School Choice. For most students with dyslexia, the margin between success and failure at college is thin, so questions of school prestige, family tradition, or social life should all take a backseat to whether a particular school will provide the necessary supports and services. When applying to schools, students with dyslexia should be honest and open with admissions officers about their needs, and they should also be critical in evaluating the responses they receive. Schools that are good environments for students with dyslexia will have a clear support system in place for these students.

Campus visits are also important to make sure that the reality on the ground matches the website rhetoric. Students with dyslexia should speak directly to staff at the disability resource center and to at least one—and preferably more—student with dyslexia who works with the center.

A good resource center should provide technology supports like text readers or recordings, help in obtaining classroom accommodations, access to notes, assistance with organization, tutors, proofing and correcting papers, advice on scheduling and instructors, and help in making contact with other students with dyslexia. Students with dyslexia should also ask about faculty attitudes toward accommodations (especially in the area of a contemplated major). If a school doesn't have a consistent and easily demonstrated record of helpfulness in all these areas, look elsewhere. Students with dyslexia almost always have a better experience at less prestigious colleges that show a greater commitment to helping them succeed than at more prestigious schools that are less committed to their support.

Helpful lists of schools with exceptional disabilities services offices are available through the American Educational Guidance Center (www.college-scholarships.com/learning_disabilities.htm). Many students with dyslexia have also found that Loren Pope's books *Colleges That Change Lives* and *Looking Beyond the Ivy League* provide helpful information about schools that are especially nurturing for students who require more personal connection to thrive in their education.[3]

An additional question for those looking to attend a four-year college is whether to attend a smaller private institution or a larger (typically state-run) university. Each can have advantages and disadvantages for students with dyslexia, and determining which provides the best fit ultimately depends upon the particular needs of the student. However, there are several typical differences that may be helpful to consider.

Larger schools often have fewer standard requirements and more options for fulfilling the requirements they do have. This greater flexibility provides students with more opportunities to select classes and instructors that better meet their needs. Larger schools also typically provide more opportunities for earning credits for independent research or practical projects, and this can allow students to avoid some coursework. Larger schools also tend to feature more courses that are geared toward providing practical career training, and this "real-world" focus can appeal to some students with dyslexia.

Smaller schools, by contrast, often provide a greater sense of community, which means students are less likely to simply become "lost." Students also typically have more direct interaction with instructors, and class sizes are often smaller. Both factors promote more back-and-forth discussion and person-to-person learning, which many students with dyslexia prefer.

Another important factor that distinguishes many large and small schools is the kinds of tests and assignments they give. Larger schools typically have large classes, which generally means more standardized tests (often multiple choice) and fewer essays or papers. Smaller schools tend to show the opposite pattern, requiring more written work. Either format may be preferred by different students, but knowing which format is the best fit for a particular student can be helpful when selecting a school.

When to Start College. The transition to college may come right after high school, or it may be made later. Because of the late-blooming developmental pattern many individuals with dyslexia follow, some students who'll eventually thrive in college may not be ready by age eighteen. Students who don't yet show the focus, drive, or motivation to enter college at eighteen may benefit from a few years off to work, attend part-time classes, serve in the armed forces, travel, or join a service organization before entering college as a full-time student. These experiences often help to focus career plans as well as to develop maturity and character, and many students find that they enjoy and excel at work in the real world more than coursework.

For some students with dyslexia, the transition to college can be gradual rather than all at once. Many students with dyslexia find that taking a full college load right after high school is too much, but a half or three-quarters load is quite doable. Often these students find that a part-time college program that is stretched over six or seven years works quite well for them. Students who are intellectually capable of high-level work but who still struggle with the speed or organization needed for heavy course loads often do well pursuing a part-time college schedule right from the beginning.

A gradual transition from high school to four-year college may also involve taking courses at a junior college. This is an especially good option for students who have difficulty motivating themselves for courses outside their core areas of interest. Junior college allows these students to take the entry-level courses they need outside of their major in settings where the competition and grading may not be as intense and where they can take fewer courses per term. Students with dyslexia also often enjoy taking entry-level or survey-type courses in a less competitive junior college environment because these courses—with their extensive reading lists, broad but shallow approaches, and emphasis on memorizing details rather than mastering big-picture concepts—often stress their weakest skills, putting them at a particular disadvantage to the "A students" who are typically in heavy supply at four-year colleges. Many junior colleges also offer online courses that are easier to balance with work schedules.

Students with dyslexia who still need to improve their academic and organizational skills before attempting a conventional college program may benefit from attending a junior college that has special programs geared to helping students with dyslexia. The first accredited junior college established specifically to prepare students with dyslexia for advanced schooling was Landmark College in Putney, Vermont (www.landmark.edu). Landmark offers a variety of programs to prepare students with dyslexia and other learning challenges in the basic academic skills that are required for success in higher education, including reading and writing, executive function and organization, self-help skills, and the use of assistive technologies. More than half the students who enroll full-time at Landmark have previously attended other colleges, where they have struggled with their coursework. At Landmark, the focus is on building the skills these students will need to transfer back to a four-year college. Some students attend Landmark for as little as one academic term or for the summer, but many attend programs that extend for as long as two years. Often students earn transferable college credits while at Landmark.[4]

Another college that focuses exclusively on educating students with dyslexia and other learning disabilities is Beacon College in Leesburg, Florida (www.beaconcollege.edu). Like Landmark, Beacon offers two-year associate in arts degrees, but it also offers four-year bachelor of arts degrees as well.

While some students with dyslexia require extra preparation before attending college, some highly capable dyslexic students may find college to be easier and more interesting than high school, and they thrive with early entry. In general, these students are highly motivated, goal-oriented self-starters with strong self-advocacy skills and good support at home. One of the individuals we interviewed, Harvard astrophysicist Dr. Matthew Schneps, followed this path. He took advantage of a program that let New York City high school juniors matriculate at City College of New York. According to Matt, "That opened the world to me. All of a sudden I went from typical high school testing, where they're just seeing how well you can memorize things, to people caring about your ideas and how you can put them together; so getting out of the school system was very helpful for me." Many bright students with dyslexia find college easier than high school because of the greater freedom it offers to focus on areas of strength and avoid areas of weakness. Finally, many colleges provide better disability support services than many high schools, which is obviously an advantage for these students.

Getting a Formal Diagnosis before Applying to College. To receive accommodations either for admissions testing or for college courses, a student must have a formal diagnosis of a special learning challenge. This diagnosis must be made by a qualified specialist on the basis of appropriate psychoeducational testing. To receive accommodations for admissions testing, the psychoeducational evaluation must usually be performed within five years of the date of the admissions test.

Examples of the tests, diagnoses, and professionals accepted by the organizations that administer the college entrance exams are available on their websites (www.collegeboard.org and www.act.org). Typically, these organizations require an IQ test (like the WISC-IV) and an achievement test (like the WIAT-III or the WJ III). We've found that many bright students with dyslexia may not show problems with reading speed or comprehension on the simpler passages contained in standard achievement tests but will struggle to read the more advanced passages found on the SAT or ACT. In general, we've found that the Nelson-Denny Reading Test is a better predictor of such problems, and it should be used in testing all students in ninth grade and beyond. Several of the other less commonly used subtests we've found to be especially helpful in documenting the important executive function and processing speed challenges that can require extra time on admissions tests include the WJ III NU subtests on decision making, visual matching, and pair cancellation.

Establishing a Record of Accommodations in School. Before granting accommodations for the admissions tests, the organizations that administer these tests (the College Board and ACT) require documentation that the student has both needed and been given accommodations in school. Documentation must include a formally accepted plan from the student's school, approved within the last twelve months. These organizations also need to know the date at which the student first began to receive accommodations, and they will usually not respond favorably if the student has only recently begun to receive accommodations. That's why it's so important to get a diagnosis and a plan of accommodations established early.

Choosing the Right College Admissions Test. The two primary admissions tests are the SAT and the ACT. The SAT, which is offered by the College Board (www.sat.collegeboard.com), is designed to assess general college readiness, while the ACT (www.act.org) tries to identify the level of academic achievement the student has reached in specific subjects, including English, reading, math, and science reasoning. Some students with dyslexia prefer one format over the other, but we haven't found a firm rule for predicting which test will appeal to which students. We would recommend that students explore sample problems from each kind of test to see which format they prefer, either on the College Board or ACT websites or using commercially available test preparation books.

Admissions Test Accommodations. Depending on the precise nature of the challenges that a student with dyslexia shows, different accommodations may be required during the admissions test session. Based on a special review, the organizations that administer the tests can provide students with accommodations that include extended time (typically double time); reader and scribe (assistance with reading and recording answers); permission to mark answers in the test booklet rather than on an answer sheet; a small room for the test session; extended or more frequent break times; permission to use the computer for writing essays; and large print on the test booklet or answer sheet. The particular accommodations granted will depend upon the documentation the student provides, including the kind of evaluation we discussed above. For students who need accommodations for testing, the testing organizations recommend applying at least nine months in advance of the desired test date. Some students may also need to complete an appeals process to receive accommodations if their initial request is turned down. This is particularly true for homeschooled students, who typically lack a formal record of classroom accommodations. For students who feel they've been unfairly denied accommodations, attorneys are available who can help families negotiate this process. Families can also check at www.wrightslaw.com for information about rights and legal recourse.

Test Practice and Preparation. Students who are unfamiliar with test procedures and exam question formats may experience confusion and make mistakes during the high-pressure testing situation, so students with dyslexia often benefit from taking practice tests in advance. Learning appropriate testing strategies and becoming familiar with the types of questions asked (and the types of words contained in the different questions) are usually well worth the effort. Because these admissions tests can be retaken, it's best not to wait until the last available date to take them for the first time, so that if necessary there will be enough time to sit for them again.

Thriving in College as a Student with Dyslexia

College presents special challenges for students with dyslexia. Factors that can make college many times more challenging than high school include:

- Increased quantities of reading and writing
- Higher standards for written work
- The loss of daily support from family and friends
- The transition from familiar settings (where many of the requirements and procedures have been mastered) to new and unfamiliar environments with equally new and unfamiliar demands
- The often enriched selection of fellow students, who are the best students from many high schools

Many students with dyslexia find the first two years of college the most difficult. During these years they must often take classes that don't match well with their strengths or interests, for reasons we described above.

The flip side is that most students with dyslexia find that when they get through the first two years of college and into their major focus, they begin to do much better. That's why students with dyslexia must do everything in their power to avoid letting 100- and 200-level courses—especially in areas outside their major—scuttle their chances to reach the stages of education where they're most likely to shine.

Selecting Courses. Careful course selection is the first key to surviving the early years of college. Students with dyslexia should carefully investigate all potential courses in advance to ensure a good fit. The better the course matches their strengths and interests, the greater their likelihood of success.

The first important factor to keep in mind is *the nature of the subject.* Is it interesting enough to provoke the necessary motivation? Is it well suited to the student's thinking strengths—including the student's MIND strengths? Does it focus on big-picture information or fine details? Does it stress the comprehension and application of concepts or principles, or the memorization and regurgitation of rote facts? Does it present information as an unfolding story or as part of a large interconnected system, or does it simply present lots of pieces of data with little attempt to tie them together?

As we've mentioned, the brains of individuals with dyslexia are often specialized to excel in big-picture reasoning rather than mastery of fine details, so looking for subjects that stress general concepts rather than precise details or memorizing minutiae can be important. For example, micro- and macroeconomics differ profoundly on this point, as do history classes that accentuate dates, times, treaties, names, etc., and those that stress the theory of history or political science. Also, inorganic chemistry—which requires a lot of bookkeeping to follow the recombination of elements and nuclei—differs greatly from organic chemistry, which leans heavily upon spatial reasoning. Students with dyslexia may also excel in courses that study complex, dynamic systems, as they involve a great deal of feedback, ambiguity, and uncertainty and require problem solving, prediction, and best-fit reasoning.

Even when subjects are a good fit, the instructor's approach may make certain courses unsuitable for students with dyslexia. Some college teachers can turn an apparently big-picture subject into a fine-detail course, so students should investigate whether an instructor has taught a particular course before and, if so, what her or his approach to teaching is. Students can also check out the ratings and descriptions of many professors at websites like Rate My Professors (www.ratemyprofessors.com).

A student's investigation into previous courses can also provide information about *the clarity and organization of the instructor*, which is a second key factor. A course may seem to match a student's interests and aptitudes well but still be a poor fit if the instructor presents the information in a way that doesn't match the student's needs. When students struggle with organization, taking notes, or learning from lectures, then they have to rely on the clarity and organization of the instructor to ensure that they'll have access to all the important information.

This touches on the third key factor, which is *the format in which the important information is presented.* This includes the nature of the reading list (e.g., how many books or pages are required for a course, what style and how clearly written the works are, and whether—if necessary—they are available on tape or digitally formatted for text-to-speech players) and the availability of resources like syllabi, note services, podcasts, or videos that contain all the information on which the student will be assessed. For students who struggle with taking notes or with auditory attention, these are especially important considerations.

A fourth key factor is *the manner in which students are assessed and graded.*

Finding courses with an assessment format that matches students' strengths is critical. Students differ in their preference for oral presentations, practical projects, written papers, multiple-choice tests, essay exams, or classroom participation and discussion. Students can master information but still fail a course if the assessment mechanism doesn't allow them to express what they know.

A fifth factor is *the instructor's attitude toward accommodations and dyslexia*. Checking with the student disabilities office and speaking directly with the prospective instructors can provide a good feel for how an instructor regards and treats students with dyslexia. Unfortunately, instructors can differ dramatically in this regard.

The sixth and final factor is *your personal passion for the subject*. The added motivation that comes from focusing on a topic you find stimulating and enjoyable is essential when college work gets challenging. When you truly love a topic so much that you can give yourself wholly to it, it can also change you in wonderful ways. A clear example of how such a passionate commitment to a single subject transformed a struggling student is provided by speech professor Duane Smith.

"Prior to discovering speech and public speaking competitions, I'd never focused on anything in my life—because I'd never had a reason to focus on anything. This lack of mental focus was apparent in my writing. When I was first starting to enjoy some success on the speech team, I gave a first draft of a speech to the team's director, and I was shocked when he chewed me out. He told me, 'There's not a complete sentence in this speech, and there's no structure. You might look good in front of an audience, but that will only get you so far if you can't structure your thoughts properly on paper.' He was very hard on me because he saw that I was trying to get by on style and wasn't working hard, but he knew that I wouldn't get much farther without really focusing and buckling down to work. That made a huge impression on me because several universities had offered me scholarships, but my coach said, 'If those schools knew your GPA, they wouldn't be talking about scholarships. You need at least a 3.0.' Now, I'd been an academic failure my whole life, and I *really* wanted to be able to tell myself that I got a full academic scholarship—and I *really* wanted to be able to tell that to my parents—so I thought, 'Wow, I'd better start going to class and focusing on things and doing what the teachers say.' It was a struggle, but I did it. And I raised my GPA, and I ended up getting a full scholarship for public speaking.

"So in the end, it was really public speaking that made me organized. The structure and discipline demanded in competitive speaking were applicable to everything else in life."

Taking on Too Much: The Risks of Overcompensation. One problem that many students with dyslexia succumb to in college is overcompensation. Blake Charlton described the pattern he fell into in college. "When I got to Yale, I was sure I didn't belong there and that I was going to fail out, so I overcompensated in spades. I almost never left the library, and I was in a regular state of panic. Even when I was elected to Phi Beta Kappa—actually, even after they gave me the diploma—I still had dreams for a long time that there was another test I'd forgotten."

Ben Foss described a similar pattern of overcompensation, but in his case it led to overcommitment outside the classroom. "In college I struggled so badly at the core academic activities, which are reading and writing, that I needed other places to be successful. So I ended up on the student affairs committee and the student magazine and in an after-school program for kids and doing fund-raising for the AIDS organization; and while from the outside I looked like 'Mr. Man-about-Campus,' I was really feeling isolated and lonely, and I was using all these activities as a way to replenish my self-image. With dyslexia there's such a thirst because all through your early years of school it's like you're in the desert and you're parched with your thirst for success, while everyone else is walking around with CamelBaks of Gatorade in terms of praise. So when you finally get to a place where you can actually get some water, you just drink and drink and drink to the point where you may actually drink more than you really should.

"Even after college I did the same thing: I did a joint law and business degree at Stanford, and from the outside it was like, 'Wow, a J.D./M.B.A. from

Stanford—you must have had no problem'; but really just the opposite was true. I had such a problem with the first year of law school that it literally put me in the hospital. I was carrying around my tape player and my books, and I was sitting in the library fourteen hours a day trying to keep up, and it literally broke me. I ruptured my 4-5 lumbar disc, but even then I kept going to class and I just didn't stop because I had this mentality that you just don't quit. I ended up working myself into the hospital and suffering so much nerve damage that I still can't feel my left foot. So in the end the answer can't just be working harder; it has to be working smarter. Like a guy in a wheelchair, you can't be dragging yourself upstairs by the elbows; you need to be using the ramp."

If at First You Don't Succeed. Many students with dyslexia who eventually succeed in college struggle at first, so by no means does initial failure equal ultimate failure. Many highly successful individuals with dyslexia succeeded in college only after several attempts or after transferring to another school that better fit their needs. Many students also do better if they take some time off coming out of high school to mature or further prepare. Ultimately, if an individual with dyslexia is committed to earning a college degree, there's no reason why—with careful planning, strategy, and persistence—this should remain beyond reach.

Summary for Students in High School and College

- One of the most important things students with dyslexia can do to prepare themselves for college is to identify their ideal learning style, which consists of their best methods of information input, information output, memory, and attention. By funneling information toward areas of cognitive strength and away from areas of weakness, students with dyslexia can learn and express knowledge as efficiently and effectively as possible.
- During high school, college-bound students with dyslexia should also:

—Develop a practical plan for meeting the reading and writing demands they'll encounter in college
—Learn strategies for organization and time management
—Develop a support network of friendships with other dyslexic students

- A college should be selected on the basis of how helpful it'll be to the student in reaching future goals and how committed and well equipped it is to help students with dyslexia succeed.
- Students with dyslexia who wish to apply to college should start early to obtain a formal diagnosis, establish a record of accommodations in high school, and apply for accommodations for testing from the admissions testing services.
- College applicants may usually take either the SAT or the ACT, and students with dyslexia should study both types of tests to see which one they prefer and to become familiar with the style of questions asked.
- Once in college, students with dyslexia should take great care in selecting courses, avoid overcommitting themselves, and make ample use of the student disabilities office.
- Students who struggle in their initial attempt at college should never conclude that they're not "college material." They should analyze the reason for their difficulties, make careful attempts to improve their weaknesses, adopt appropriate strategies, and return to school at a later point if it's necessary to pursue their dreams.

CHAPTER 29

Thriving in the Workplace

For individuals with dyslexia, opportunities to thrive in the workplace are now more plentiful than ever. Advances in technology have made it easier for individuals with dyslexia to access printed information, express their ideas in writing, and remain organized and on schedule. With self-understanding, self-advocacy, support, persistence, and careful planning, individuals with dyslexia can confidently pursue any occupation for which their interests and abilities otherwise suit them.

In this chapter we'll focus on three things that individuals with dyslexia should do to maximize their chances of thriving in the workplace: find a job that fits them well, take steps to make that job fit even better, and obtain the support and advice of other individuals with dyslexia. We'll also discuss the difficult issue of whether or not to disclose one's dyslexia to employers or co-workers.

Jobs That Fit Well

For individuals with dyslexia, good-fitting jobs have several common features. First, they *engage strengths* and *avoid weaknesses.* As we've discussed, many individuals with dyslexia excel in big-picture reasoning, or the ability to see the overall features, "contours," or implications of objects or ideas. The occupation or position in which they best display this ability depends upon which MIND strengths they possess, but as a general rule, jobs that fit individuals with dyslexia well stress problem solving, troubleshooting, fixing things, coming up with new ideas, thinking about what's missing or not being addressed, or telling stories (e.g., sales, counseling, coaching, advertising, entrepreneurship). We list the kinds of jobs that often are a good fit for individuals with dyslexia in Appendix B.

In contrast, individuals with dyslexia often struggle with fine-detail processing, mastering routine procedures to the point of automaticity, or rote memory. As a result, they often find that jobs that stress repetition, efficiency, consistency, attention to details, use of procedures, application of fixed rules, or routine processing tasks (especially clerical tasks that involve the manipulation and use of written symbols) are a poor fit.

Individuals with dyslexia may also have other strengths that they can use to find a job that fits them well. Although they may struggle to build academic credentials, their practical, "real-world" strengths can help them build other credentials, like a portfolio of work, a list of references, or a set of professional or personal connections. These skills and experiences can be gathered while working at entry-level jobs, in the military, or with volunteer organizations. Individuals with a strong interest in a particular field can also volunteer their services to a potential employer for a trial period or internship, then work their way into permanent employment by demonstrating skills on the job. Personal connections, like introductions from previous teachers or employers, or even family businesses or contacts, can also be helpful in finding a good job.

A second key feature of jobs that fit individuals with dyslexia well is that they *engage interests.* While everyone works better on tasks they find interesting and enjoyable, individuals with dyslexia are often especially dependent upon interest to produce their best efforts. In contrast, when tasks fail to engage their interest, they often struggle to perform well and remain focused. This is largely because many of the rote or automatic skills needed to perform routine tasks require more focused attention for individuals with dyslexia. This need for heightened attention can be difficult to sustain unless there are things about the job that are especially interesting. When work heightens interest and mood, dyslexics typically respond with greater creativity and performance.

Job-related interest can be of two types: intrinsic interest, or interest in the job itself, and extrinsic interest, or interest in the things the job can lead to, like money, status, or reaching personal goals. Either kind of interest can help to build focus and engagement in a job. Ideally, individuals with dyslexia should look for jobs that combine both kinds of interests. As entrepreneur, billionaire, and dyslexic Richard Branson observed of his own career, "I set out to create something I enjoyed that paid the bills." The results, to put it mildly, speak for themselves.

A third key feature of jobs that fit individuals with dyslexia well is that they *focus on results rather than on methods.* Many of our interviewees mentioned that they often perform tasks in unconventional ways—frequently of their own devising. For example, more than half told us that they solved math problems differently from how they were taught by using unconventional methods that made more sense to them. This is also a frequent finding among the individuals with dyslexia that we see in our clinic. A preference for doing things atypically is one key reason that jobs that stress uniformity of method are often a poor fit for dyslexics. (Recall Sarah Andrews struggling with her boss, who expected her to perform creative work using creativity-stifling methods.)

In contrast, jobs that allow flexibility can open the door to success for dyslexics. It's often while devising new methods for routine tasks that dyslexics come up with innovative approaches that save time, effort, and expense and improve outcomes for everyone. Learning expert Dr. Angela Fawcett commented, "I think one of the benefits that dyslexics experience from their difficulty mastering procedures is that they have to rethink tasks each time, starting from fundamental principles, instead of having the steps entirely automated and ready to be performed without thinking. Because they can't rely on these automatic skills in a mindless fashion, they're not limited by the rules, so they can think *outside* the rules. I think this helps them think more creatively than if they were, in a sense, trapped within the rules."

Similarly, individuals with dyslexia often find that jobs with flexible task assignments fit better than those with more rigid and fixed task assignments, because they allow them to focus on the things they do best. Sometimes smaller or newer companies offer greater flexibility. Disabilities rights advocate Ben Foss told us, "In general, large corporations deal with consistency, but small corporations deal with variations, so it can be harder to break the norms in larger corporations. You can sometimes find flexibility in a larger company, too, but only if that company has a system in place to deal with variation, which allows it to be flexible and adjust to people's needs."

There is evidence that this kind of flexibility is often more easily found in positions very near the top or the bottom of the structures of large organizations but in shorter supply in the middle. Professor Julie Logan has found that although many large corporations have CEOs with dyslexia, fewer than 1 percent of middle managers in such firms are dyslexic.

This doesn't mean that it's impossible for individuals with dyslexia to advance up the ladder in a corporate system, but it does mean that they are less likely to do well in midlevel corporate positions unless the companies they work for take special care to make their jobs a good fit. Ben Foss stressed that each potential employer must be evaluated individually regarding its ability to provide the flexibility necessary for an individual with dyslexia to survive in its corporate structure. Some large companies, like his former employer Intel, manage to maintain their flexible attitudes despite their size. Douglas Merrill also told us that supporting this diversity in thinking styles was one of his primary goals as chief information officer at Google. Douglas worked hard to give employees the greatest possible flexibility in choosing the work habits and technologies that allowed them to be their most productive. When a company shows this kind of flexibility, it's likely to be a good fit for individuals with dyslexia.

Of course, there's no employer that can provide more flexibility than oneself, which is one reason why so many dyslexics start their own businesses. But with regard to working with established companies, we've observed in general that individuals with dyslexia experience more success with smaller, younger, more flexible, and more creative companies.

Steps to Improve Job Fit

After choosing a job that seems to be a good fit, individuals with dyslexia should work hard to optimize that job environment by *being proactive* in pursuing opportunities, *self-advocating* with supervisors and co-workers, *building partnerships, pursuing leadership opportunities*, and *using technologies* to maximize their productivity.

Many individuals with dyslexia are especially good at spotting opportunities that others have missed and then aggressively and proactively taking advantage of those opportunities. Professor Julie Logan cited this ability as one of the most common characteristics she's observed in the dyslexic entrepreneurs she's studied.

We've also observed this ability in many of the individuals with dyslexia we've interviewed—and not just in business. Astrophysicist Matt Schneps told us, "One thing I'm very proud of is that I'm very good at taking advantage of opportunities. If I see something I think is useful for me, I think about how I can make the most of it and take advantage of that." Because of this ability (and strong self-advocacy skills like those we'll discuss later), Matt has been able to enjoy four entirely different careers over the past thirty years, all with the same employer.

Author Vince Flynn provides another great example of how individuals with dyslexia can find and aggressively pursue unusual pathways to success. He was so convinced that his first novel, *Term Limits*, would appeal to a large reading audience that even though he'd received over sixty rejections from publishers, Vince decided to self-publish two thousand copies, then sold them all himself at a booth in a shopping mall. Only after this put him on a local bestseller list did Vince land a contract with a big New York publisher. Not long after, *Term Limits* went on to become the first of Vince's twelve consecutive bestsellers.

When individuals with dyslexia emerge from school with an appropriately positive view of themselves and of their futures, these individuals are often remarkably resilient and confident in their abilities to achieve what they set out to do. Professor Julie Logan commented on these traits in the entrepreneurs with dyslexia she's studied. "Learning to cope with and solve all the problems they face in school gives many dyslexics this 'can-do' approach that they bring to all sorts of new situations. They just know they can make things work."

However, before individuals with dyslexia are given the chance to prove that they can accomplish things, they must often convince others to give them an opportunity—and this requires self-advocacy. Self-advocacy is the ability to say, "Here's what I do well—here's how I think I can best contribute." Self-advocacy involves persuasion, negotiation, and the ability to tell a convincing story about yourself. Many individuals with dyslexia have been advocating for themselves since their earliest days in school, so they often excel in self-advocacy—frequently without realizing it. Professor Logan commented, "I think many dyslexics aren't aware of how good their people skills are. Even the dyslexics in our mentoring program [which we'll discuss in a moment] who haven't been all that successful in their careers are often really good at getting their mentors to do things for them—even when the mentor wasn't expecting to. They often have this amazing ability to get others to do things for them. That's something they should really learn to use."

In fact, many of the individuals we interviewed emphasized the importance of learning to work well with partners or teams. Douglas Merrill told us, "It's really critical to surround yourself with diversity. There's lots of evidence—including research I did—that diverse teams tend to yield better outcomes." But he also stated that forming good working relationships isn't always easy, and it requires careful communication. "At RAND there was a researcher who wasn't so strong with long-term vision but was really outstanding at working with details, and because we were good at different things I tried to build an ongoing research relationship with him. Unfortunately, I ended up offending him because I wasn't good at describing the positive things he did; so he sort of heard me saying, 'I'm smarter than you,' which really wasn't what I was trying to say. It's important to make sure when you're working with partners that they hear you saying, 'We're good at different things, and that's fine.'"

When appropriate, individuals with dyslexia should also be willing to take a leading role in groups and partnerships. Professor Logan found that many of the dyslexic entrepreneurs with whom she works possess interpersonal skills that make them particularly well suited to leading groups. "Business is actually mostly about managing people, and managing teams, and dealing with the unexpected, and those are things dyslexics do every day of their lives. They're always having to negotiate their way around difficulties, or having to make new plans because they lose things, or going to the appointment on the wrong day or turning up at the wrong time. So over time they develop a lot of skills, like the abilities to delegate, manage, and inspire people—all of which are important leadership skills."

Finally, individuals with dyslexia can optimize their fit at work by using technologies to increase their productivity. Of particular value are devices that help them with organization, reading (text-to-speech), and writing (speech-to-text and special dyslexia-friendly word processing programs). We list these devices in Appendix A.

Getting Support and Advice from Other Individuals with Dyslexia

Relationships with dyslexic peers can also be tremendously helpful for individuals with dyslexia. There's a large body of research documenting the feelings of inferiority and isolation dyslexic individuals often experience due to their challenges in school. These feelings typically persist as long as individuals with dyslexia remain cut off from others with similar experiences.

Ben Foss shared the following story from his experience at Intel. "A graphic designer came to me when we had just started working on the Intel Reader [a portable text-to-speech reader], and he said to me very secretively,

'I just want to tell you—and you can't repeat this to anybody—but I'm dyslexic.' Now, I found this to be absolutely mind-blowing because this designer is paraplegic, and he's been in a wheelchair for the last twenty years. Yet the reason he was afraid that other people might think he couldn't do his job well was because of his dyslexia rather than his paraplegia. When I talked with him about it, it turned out that he had failed third grade because of his dyslexia, and he was still incredibly embarrassed about that, despite the fact that he was a very successful designer at Intel. Fortunately, having the chance to talk this over with a dyslexic co-worker was really helpful for him. However, I cannot overemphasize how strong this feeling of isolation can be for individuals with dyslexia."

Because these feelings of isolation can be so severe, relationships with other individuals with dyslexia can be extremely liberating and empowering. As Ben put it, "When you meet another dyslexic, it's like you're immigrants meeting in a new land. You instantly know important things about each other, and that experience of being from the same country is incredibly powerful."

Ben's vision for building community among dyslexics has led him to found a peer membership organization called Headstrong Nation (www.headstrongnation.org). Ben explains, "We need someplace beside the special ed trailer out back of the school where dyslexics can meet to congregate. The underlying sense of isolation, and the need for information about learning accommodations and practical advice for work—all of those are shared experiences, so why don't we get together and share them? That's what Headstrong is trying to do."

This vision of bringing dyslexic individuals together is one we deeply share. At our Dyslexic Advantage website, individuals with dyslexia—as well as their families, friends, and the professionals who care for them—can all find important information and a sense of community. On the site we've posted video interviews with many remarkable individuals with dyslexia; reviews of curricula, schools, educational software, and technology devices; discussion forums for dyslexic individuals in different professions; and further resources for parents, educators, and others.

Another valuable form of support for dyslexic individuals is the advice and guidance of other dyslexic individuals who've struggled—and succeeded—at similar jobs. This is true not only at the start of one's career but at any point where difficulties are encountered. For the last two years Professor Julie Logan—along with the British Dyslexia Association, Dyslexia Scotland, and the Cass Business School—has been involved in establishing a mentorship program in the United Kingdom for working adults with dyslexia. Dr. Logan described this program to us:

"Basically, we pair a very successful dyslexic who's already established in his or her career—and who's been through the same problems—with a dyslexic who's in need of support. Initially, we expected that most of the dyslexic individuals asking for mentors would be in their twenties—in their first job or just coming out of university. But the average applicant turned out to be much, much older—almost twice what we expected. One reason for this is that in the United Kingdom, at least, people are finding out they're dyslexic quite late in life; so they're joining our program because they realize, 'Now I know what my problem is, perhaps I can get help, and perhaps I can get on in my career.'

"Often, they've been struggling because they keep making the same mistakes again and again. For example, one woman was going through this cycle: because she had low self-esteem, she would push herself really, really hard; then she would get promoted, but she would quit almost immediately because she couldn't cope with all the new responsibilities that came with the promotion. Like a lot of our mentees, she had this tremendous desire to prove that she wasn't sick or stupid, and that she could do everything, but it worked to her detriment. What we've been able to do for her is put her with a mentor who understands what she's going through from personal experience. That mentor has taught her strategies, and how to pace herself, and how not to push herself so hard, and how to value what she's achieved so far. As a result of that mentorship, she's been promoted to a customer account manager position, and she's been able to cope."[1]

Disclosing Your Dyslexia: Yes or No?

A final issue that all individuals with dyslexia must deal with is whether or not to disclose their dyslexia to employers and co-workers, either at the time of the initial interview or on the job. This question often provokes strong opinions from advocates on both sides of the issue, and there's probably not one answer that's right for all individuals in all jobs. One thing, however, is absolutely clear: by law, disclosure is not required, so all individuals with dyslexia can decide for themselves whether or not to disclose. The following considerations may be helpful in making this decision.

In an ideal world, individuals with dyslexia would never need to conceal their dyslexia, because their employers and co-workers would understand what we've explained in this book: that dyslexia is associated with strengths as well as challenges, and that individuals with dyslexia have abilities that can be useful in almost any kind of job. Unfortunately, our world hasn't yet reached that ideal state, and ignorance keeps some employers leery of giving individuals with dyslexia a chance to prove their worth. That's why individuals with dyslexia have traditionally been counseled *not* to disclose their dyslexia in job interviews—especially when seeking an entry-level position or a first job. Professor Julie Logan expressed this view when she told us, "Letting people know that you're dyslexic doesn't always work for everyone—especially people who are employed by others—because there's still a lot of prejudice about dyslexia."

Ben Foss shared a different opinion. "We can get fixated on this question of disclosure, but the question is really context: should you provide people with context for understanding your dyslexia? And the answer is almost always yes. In general, it's much better to know right up front that a potential employer is a bigot, so you can go somewhere else. Otherwise, you'll have to deal with that prejudice later, after you've already accepted the job—and that usually ends badly."

Ben's preference for disclosure was formed when he was in professional school. "My own sense of strength and comfort with my dyslexia came through meeting other people with disabilities—mostly people with physical disabilities. At Stanford Business School I had a remarkable classmate named Mark Breimhorst. Mark was born with no hands, and he's also paralyzed on the left side of his face, so he can't smile and he can't wink, and Mark understood that those things are unnerving to some people. So during the first week of business school, Mark sent out an e-mail to everyone in which he provided context for how to deal with him. He wrote: 'My name is Mark. When you meet me, you'll see that I have no hands. If I extend my arm, shake my wrist. When class is over, I don't need your help carrying my bag out: I brought it in.' And in all these ways Mark gave us context for how to interact with him, and that made everything so much easier.

"Mark also helped me see how these kinds of steps could be helpful for someone with dyslexia. When Mark found out that I was dyslexic, he invited me to be on a panel for our classmates to discuss disabilities. At first I felt self-conscious because I didn't have a physical disability, and I'd never talked about my dyslexia in public before. But Mark was adamant. His view was that we were all going to be future leaders, so we should help our colleagues learn what disabilities are all about, so they'd have the right context for dealing with other individuals who are different. And he was right. I learned a tremendous amount from Mark about how to deal appropriately with my challenges. By the time I left school and went to work for Intel, it was easy for me to say right up front, 'I'm dyslexic. Here is some context on me and this is what I can do well.'"

When enough individuals with dyslexia are willing, like Ben, to step out and show skeptics how capable dyslexics can be, it will become easier for all dyslexics to be open about their challenges in the workplace. But for now, the decision to inform others about one's dyslexia must remain a personal one.

While disclosure is often difficult for individuals with dyslexia who are employed by other people, for individuals who are self-employed the risk/reward calculation often leans more toward disclosure. According to Julie Logan, "When people are their own bosses, letting other people know that they're dyslexic isn't usually a problem. Many of the dyslexic entrepreneurs with whom we work are very open about the fact that they can't spell, for example, and I think those sorts of disclosures help, because by being quite open about their issues and weaknesses, they actually make people want to help them. For example, when they send their e-mails, if they tell people they're dyslexic and that's the source of whatever errors there may be in their writing, the recipients don't worry that their business skills are similarly haphazard and start condemning their business persona. One of our very successful dyslexics recently sent out an e-mail that could have been taken as quite rude because of an error she made, but fortunately at the end of all her e-mails there's a little caveat in her signature that says, 'These e-mails won't be spelled right because I'm dyslexic.' That sort of disclosure is often quite helpful."

Summary for Individuals with Dyslexia in the Workplace

- Today, opportunities for success in the workplace are greater than ever before for individuals with dyslexia, and they should feel free to pursue any job for which they are otherwise suited by their interests and talents.
- Keys to succeeding in the workplace include finding a job that's a good fit, taking steps to make that fit better, and getting the support and advice of other individuals with dyslexia.
- Jobs and careers that are great fits for individuals with dyslexia capitalize on strengths, avoid weaknesses, engage interests, and focus on results rather than on methods.
- Steps to make jobs a better fit include being proactive in pursuing opportunities, self-advocating, forming partnerships, pursuing opportunities for leadership, and using technologies to enhance productivity.
- Forming relationships with peers and mentors with dyslexia can be invaluable for providing emotional support and practical advice.
- The decision to inform employers or co-workers about dyslexic challenges is legally yours to make. Good arguments can be made for either disclosure or discretion, but each individual should decide in advance which choice is right for her or him.

Epilogue

Throughout this book, we've tried to answer the question, What does it really mean to "be dyslexic"?

As we've explored this question, we've learned that "being dyslexic" involves much more than having "a disorder of reading and spelling." Ultimately, being dyslexic means that you have a nervous system that's built to work differently—and it's been built to work differently because there are remarkable advantages to having a brain that works in this way.

Early in this book we used several metaphors to illustrate the relationship between strengths and challenges in dyslexia. We'd now like to share a final metaphor that illustrates our view of dyslexia as a very different, but equally valuable, way of processing information.

Imagine that you arrive for an appointment at a nicely furnished office and are asked to wait for a few moments. As you take a seat, you notice a variety of knickknacks on the table beside you. Curious, you begin to look them over, and your eye is caught by a long, thin, glasslike rod, triangular across its short axis. You pick it up, examine it, and notice that it's translucent. You wonder if it's a lens of some sort—perhaps a magnifying glass—so you hold it closer and try to look through it. But no matter how you manipulate it, it fails to improve your vision. Eventually you grow frustrated with this seemingly worthless piece of glass; but as you reach to set it down, it passes through a beam of light from the office window. Suddenly, a flash of brilliant rainbow colors appears on the surface of the table beneath the rod. In that instant, you realize that the rod in your hand isn't a defective lens but a perfectly functioning prism.

Like this prism, the dyslexic mind has provoked attention and interest, but its true nature and purpose have been missed. It's been evaluated for its clarity and accuracy as a lens, and found wanting. Yet if we study the dyslexic mind carefully, we'll find that its true excellence is its ability to reveal many things that are hard for "normal" minds to see.

The true importance of the dyslexic mind lies in its MIND strengths. While not every individual with dyslexia enjoys all the MIND strengths, most show one or more of these important abilities, and these strengths often provide the keys to their success.

In emphasizing these strengths, we have no wish to downplay the very real challenges that individuals with dyslexia face—and in many cases, the actual suffering they experience. But as we said at the beginning of this book, "suffering from dyslexia" is suffering of a very special kind. Rather than the suffering of a person with an incurable disease, it's the suffering of a hero on a perilous but promising quest.

Perhaps the hero who best represents this dyslexic quest is Aragorn, from J. R. R. Tolkien's *The Lord of the Rings.* Early in the Ring saga, Aragorn seems little more than a wandering vagrant. Yet he is the rightful heir to the throne of Gondor and destined one day to be king. Aragorn's fate is foretold in the lines of an ancient prophecy, which reminds us that all things are not as they first appear, that royal natures are sometimes hidden beneath rags:

All that is gold does not glitter,
Not all those who wander are lost;
The old that is strong does not wither,
Deep roots are not reached by the frost.
From the ashes a fire shall be woken,
A light from the shadows shall spring;

Renewed shall be blade that was broken,
The crownless again shall be king.

If you are an individual with dyslexia, this prophecy is also for you. Although you may not "glitter" in the classroom, if the promise of your future can sustain you through the challenges you face, your intellect will be forged as keen as any blade.

The truth of this "prophecy" doesn't rest on the words of an ancient oracle but on scientific research, clinical observations, and the experiences of the countless talented and successful individuals with dyslexia who've come before you. The information we've presented in this book allows us to say with complete confidence that:

- if you persist through hard times when further effort seems pointless;
- if you work diligently on proven therapies;
- if you use helpful technologies and accommodations;
- if you pursue support and guidance from other individuals with dyslexia;
- if you set goals for your future, become a skilled self-advocate, and proactively pursue opportunities;
- if you try your hardest each day without worrying whether you see obvious progress;
- if you never stop believing in the certainty of your positive future; and
- if you make full use of the strengths that you possess;

then in the end you'll find that rather than being "cured of dyslexia," you'll become a perfect example of what an individual with dyslexia was always meant to be.

And when you do, you'll understand the truth, and the true nature, of the Dyslexic Advantage.

APPENDIX A

Accommodations and Resources

I. School Accommodations for Reading and Writing

- Students with significant delays in reading development and poor reading fluency and/or comprehension should be allowed and encouraged to use *recorded books* and *text-to-speech software* as soon and as often as possible for all assignments that involve learning from or interacting with text (e.g., filling out worksheets, answering test questions). For such students, reading should be thought of as an activity to be practiced for its own sake, rather than as a means to other kinds of learning.
- Students with significant problems with reading fluency will require *extra time* for in-class tests and assignments.
- Students with significant problems with reading speed and/or comprehension will typically also require either *oral testing* or a person to act as a *reader* to read the test questions aloud to them.
- For some students with reading problems, especially in the early grades, access to *large-print books* can be helpful.
- For students who struggle with accurate decoding and/or word

retrieval, the use of a *talking dictionary* and *electronic thesaurus* that will pronounce and define words that have been entered into it can be useful for identifying words that can't be easily guessed and for finding alternative words.

- Dyslexic students should be allowed the widest possible latitude in *choosing reading materials* when the goal is reading practice. When the goal is building exposure to literate information, dyslexic students should be given flexibility to pursue such information through *listening* rather than through reading.
- *Extra time* for handwritten assignments.
- *Reduced quantity of written assignments* for students struggling to complete work (worksheets, spelling lists, essays, reports, math, etc.).
- *Correction without point deductions* for mistakes in spelling and mechanics.
- Permission to use a *keyboard* for all assignments longer than single words (or in some cases sentences).
- Permission to *orally dictate* longer assignments if keyboarding fluency is not yet achieved.
- A *scribe* for tests requiring extensive writing or filling in bubbles, or an option for oral testing.
- Permission to use *readers to help correct papers* before turning them in.
- Chances to *rewrite* or correct mistakes on assignments.
- Provision of teacher's *notes*, or notes copied from another student for lectures.
- Permission to use a *keyboard and/or recording system* in class.

II. Resources for Reading

Resources for training in phonological skills and phonics

- International Dyslexia Association (www.interdys.org)
- Lindamood-Bell Learning Centers (www.lindamoodbell.com)
- Institute for Excellence in Writing (www.excellenceinwriting.com)

- Starfall.com (www.starfall.com)
- Headsprout (http://headsprout.com)

Computer-based auditory training

- Earobics (www.earobics.com)
- Fast ForWord (www.scientificlearning.com)

Recorded books

- RFB&D (Recording for the Blind and Dyslexic) (www.rfbd.org)

Digital text repositories

- Bookshare.org (also provides reader software)
- For classic texts: Project Gutenberg (www.gutenberg.org)
- Research for articles, magazines: Questia.com
- Commercial readers, like Amazon's Kindle, Barnes & Noble's Nook, or the Sony Reader (which also include text-to-speech)

Text-to-speech technologies

- Intel Reader (www.intel.com/healthcare/reader/about.htm)
- Kurzweil 3000 text-to-speech technology (www.kurzweiledu.com/kurz3000.aspx)
- Read&Write Gold (www.texthelp.com)
- ReadingBar for Internet browsers (www.readplease.com)

III. Resources for Writing

Instructional materials

- Handwriting Without Tears (www.hwtears.com)
- *From Talking to Writing: Strategies for Scaffolding Expository Expression* by Terrill M. Jennings and Charles W. Haynes (available only at http://www.landmarkoutreach.org/pub181.htm)

- *Writing Skills 1* and *Writing Skills 2* by Diana Hanbury King
- *Step Up to Writing* (www.stepuptowriting.com)

Word processing software with spell-checking, grammar-checking, and read-aloud functions

- Read&Write Gold (www.texthelp.com)
- Ginger Software's contextual grammar and spell-checker (www.gingersoftware.com)
- Write:OutLoud (early elementary) and Co:Writer (middle elementary and above) (both available from www.donjohnston.com)

Visual planning, brainstorming, and mind-mapping software

- Inspiration (adult) and Kidspiration (child) software (www.inspiration.com)
- XMind open-source mind-mapping software (www.xmind.net)

Speech-to-text (oral dictation) software

- Dragon Naturally Speaking is a popular speech-to-text, or oral dictation, software program that allows the writer to dictate text into a computer microphone and then translates the speech into printed text on the person's computer (for word processing or e-mail) or the person's cell or smartphone. We've found this program to be an excellent aid for adults and students beyond mid adolescence, while younger students typically have difficulty making it work and do better orally dictating to a parent, tutor, or other scribe. (www.nuance.com)

Note-taking technology

- Livescribe smartpens (www.livescribe.com)

IV. Resources for Time Management and Organization

Organizational resources

- Traditional materials like message boards, sticky notes, checklists, or Day-Timers
- Vendors that send reminders about appointments or "to do" items to a cell phone or computer, like:

    —Remember the Milk (www.rememberthemilk.com)
    —Skoach (www.skoach.com)
    —Google Calendar (www.google.com)

- Electronic timers to help improve time awareness and focus during tasks:
    —On the computer: TimeLeft (www.timeleft.info)
    —On the tabletop or wristwatch: Time Timer (www.timetimer.com)
- Good source for other helpful organizational strategies: Lifehacker (www.lifehacker.com)

V. Resources Related to College

Information on colleges specifically for dyslexic students

- www.landmark.edu
- www.beaconcollege.edu

Information on services provided by different colleges for students with dyslexia

- American Educational Guidance Center (www.college-scholarships.com/learning_disabilities.htm)
- *The K&W Guide to Colleges for Students with Learning Disabilities*, 10th edition, by M. Kravets (published by the Princeton Review)

Admissions test information

- SAT: www.sat.collegeboard.com
- ACT: www.act.org

Information on legal issues relating to learning challenges and accommodations

- www.wrightslaw.com

Professor ratings

- www.ratemyprofessors.com

Mentorship organization

- Project Eye-to-Eye (www.projecteyetoeye.org)

VI. Resources for Networking and Support

Projects working to encourage the growth of a true dyslexic community

- The Dyslexic Advantage website (http://dyslexicadvantage.com) and Dyslexic Advantage on Facebook (www.facebook.com/dyslexic advantage)
- Headstrong Nation (www.headstrongnation.org)
- Project Eye-to-Eye (www.projecteyetoeye.org)
- Being Dyslexic (www.beingdyslexic.co.uk)

Additional up-to-date information and product reviews can also be obtained on our website (http://dyslexicadvantage.com).

APPENDIX B

Popular Careers for Individuals with Dyslexia

Below we've listed some of the occupations that are often good fits for individuals with dyslexia. We've grouped these occupations by dominant MIND strength, but please remember that this is meant to be only a rough guide. Most individuals with dyslexia will possess more than one MIND strength, and most of these occupations benefit from the contributions of several MIND strengths.

High M-Strength Occupations and Fields

Engineer
Mechanic
Construction (electrician, carpenter, plumber, contractor)
Mathematician
Interior designer, industrial designer
Illustrator, graphic artist, graphic designer, architectural drafting
Architect
Medicine (surgery, radiology, pathology, cardiology)
Painter
Sculptor

Photographer
Filmmaker, director
Landscaper
Sailor
Airplane pilot
Orthodontist, dentist, dental hygienist

High I-Strength Occupations and Fields

Computer or software designer (networks, programming, systems architecture)
Scientist (zoologist, biochemist, geneticist, chemist, environmental scientist, geologist, paleontologist, physicist, astronomer, astrophysicist)
Naturalist, environmentalist
Inventor
Museum director
Clothing or fashion designer, tailor, seamstress
Dancer, choreographer
Musician
Actor
Chef
History, political science, sociology, anthropology, philosophy
Comedian
Nurse
Therapist (physical, occupational, sports)
Trainer

High N-Strength Occupations and Fields

Poet, songwriter
Novelist
Literature, journalism
Screenwriter
Counseling, psychology, ministry

Coaching
Teaching
Public speaking
Politician
Game or video game designer
Attorney (especially litigation, tax law, criminal defense or prosecution, arbitration)
Sales
Marketing
Advertising
Public relations

High D-Strength Occupations and Fields

Entrepreneur
Chief executive
Finance (trader, investor, venture capitalist)
Small business owner
Business consulting
Logistics, planning
Accounting (tax planning, consulting, CFO)
Economics (especially macroeconomics)
Medicine (immunology, rheumatology, endocrinology, oncology)
Farmer, rancher

NOTES

Preface

1. http://www.cass.city.ac.uk/media/stories/story_8_45816_44300.html.

Chapter 1

1. Current research suggests that as many as 20 percent of U.S. residents may be considered dyslexic. See, e.g., S. E. Shaywitz, *Overcoming Dyslexia* (New York: Alfred A. Knopf, 2003).
2. Morgan, W. P., A case of congenital word-blindness. *British Medical Journal* 2 (1896): 1378.

Chapter 3

1. B. L. Eide and F. Eide, *The Mislabeled Child: Looking Beyond Behavior to Find the True Sources—and Solutions—for Children's Learning Challenges* (New York: Hyperion, 2006).
2. This tendency for single "upstream" variations to cause a wide variety of "downstream" effects is typical of integrated systems. Think, for example, of the electrical system in your house, which affects things as different as lighting, refrigeration, cooking, and entertainment. One "upstream" change in the electrical circuit board could cause "symptoms" in all these "downstream" functions.

3. S. Dehaene, *Reading in the Brain: The Science and Evolution of a Human Invention* (New York: Viking, 2009), and M. Wolf, *Proust and the Squid: The Story and Science of the Reading Brain* (New York: Harper Perennial, 2007).
4. The other major branch of learning and memory is called *declarative*, and it involves learning the facts about something. We discuss declarative memory in detail in chapter 16.

Chapter 4

1. Widely read examples include Daniel Pink, *A Whole New Mind: Why Right-Brainers Will Rule the Future* (New York: Riverhead, 2005), and Betty Edwards, *Drawing on the Right Side of the Brain* (New York: Jeremy P. Tarcher, 1989).
2. For those interested in a good general discussion of these differences, we recommend R. Ornstein, *The Right Mind: Making Sense of the Hemispheres* (New York: Harcourt Brace, 1997).
3. T. G. West, *In the Mind's Eye: Creative Visual Thinkers, Gifted Dyslexics, and the Rise of Visual Technologies* (Amherst, MA: Prometheus Books, 2009).
4. Tufts University professor Dr. Maryanne Wolf also comments on this apparent dyslexia/right hemisphere connection in her fascinating survey of reading and reading challenges, *Proust and the Squid: The Story and Science of the Reading Brain* (New York: Harper Perennial, 2007).
5. S. Shaywitz et al., Functional disruption in the organization of the brain for reading in dyslexia. *Proceedings of the National Academy of Sciences, USA* 95 (1998): 2636–51. This pattern has also been confirmed by many other researchers. For a detailed discussion of this reading circuit, see M. Wolf, *Proust and the Squid*, 165–97, or S. Dehaene, *Reading in the Brain*, 235–61.
6. M. Wolf, *Proust and the Squid*, 186.
7. P. Turkeltaub, L. Gareau, L. Flowers, T. Zeffiro, and G. Eden, Development of neural mechanisms for reading. *Nature Neuroscience* 6 (2003): 767–73.
8. In R. Ornstein, *The Right Mind*, 174.
9. M. Jung-Beeman, Bilateral brain processes for comprehending natural language. *TRENDS in Cognitive Sciences* 9 (2005): 512–18.
10. The actual experiment by which Beeman demonstrated this is described in the *TRENDS* paper cited above and runs as follows. First, individuals were "primed" by being shown three different words (in this example *foot*, *glass*, and *pain*) that are each distantly related to a particular word (in this case *cut*). Second, the word *cut* was shown either to the left or the right hemisphere of the brain (by display-

ing it exclusively to either the right or the left half of the visual field). When this was done, only the right hemisphere responded more strongly than it did when the "prime" was not given, because only its broader semantic field created a "cumulative priming effect" through which the activation of each related word added additional force. In contrast, when a word more closely related to *cut*—like *scissors*—was used to "prime" subjects prior to viewing cut, it was now the left hemisphere that showed a greater priming effect. In short, the right hemisphere recognizes secondary or more distant semantic relationships that help capture overall meaning or gist, while the left hemisphere recognizes almost exclusively the "tight" primary meanings that help maintain precision.

11. A useful summary of this work is provided in E. L. Williams and M. Casanova, Autism and dyslexia: A spectrum of cognitive styles as defined by minicolumnar morphometry. *Medical Hypotheses* 74 (2010): 59–62.
12. That is, autistic individuals tend to interpret messages based on a very narrow, literal, or "concrete" understanding of the words used, relying almost entirely on the primary word meanings.
13. Although many of the cognitive features associated with the bias toward long connections are similar to the features associated with increased "right-brain" processing that we described above, one advantage that Dr. Casanova's minicolumnar theory of dyslexia has over the right-brain-predominant theory that we discussed above is that it does a better job of explaining why individuals with dyslexia typically retain a "right-brain flavor" to their processing style even when brain scans show that their circuitry has become increasingly left-sided through practice. For example, we often find that individuals with dyslexia who've become relatively skilled readers still process stories in a highly gist-dependent, top-down fashion, just like many less-skilled dyslexic readers. We'll explain this finding in more detail in our section on I-strengths, but the basic point is that certain aspects of the dyslexic processing style are unlikely to completely vanish even with extensive training, as we might have predicted with the hemispheric theory, because the difference in minicolumnar orientation and bias toward long connections means that the *left* hemisphere of an individual with dyslexia will in some ways function with a rather *right* hemispheric flavor.

Chapter 6

1. E. A. Attree, M. J. Turner, N. Cowell, A virtual reality test identifies the visuospatial strengths of adolescents with dyslexia. *Cyberpsychology and Behavior* 12 (2009): 163–68.

2. C. von Károlyi, Visual-spatial strength in dyslexia: Rapid discrimination of impossible figures. *Journal of Learning Disabilities* 34 (2001): 380–91.
3. J. S. Symmes, Deficit models, spatial visualization, and reading disability. *Annals of Dyslexia* 22 (1971): 54–68.
4. N. Geschwind, Why Orton was right. *Annals of Dyslexia* 32 (1982): 13–30.
5. Psychologist Alexander Bannatyne noted that in his experience "parents in highly spatial occupations, such as surgeons, mechanics, dentists, architects, engineers and farmers, tend to have more dyslexic children than do those in other occupations." A. Bannatyne, *Language, Reading and Learning Disabilities: Psychology, Neuropsychology, Diagnosis and Remediation* (Springfield, IL: Charles C. Thomas, 1971).
6. http://www.timeshighereducation.co.uk/story.asp?storyCode=155324§ion code=26; http://www.hhc.rca.ac.uk/resources/publications/CaseStudies/id4307 .pdf.
7. B. Steffert, Visual spatial ability and dyslexia. In *Visual Spatial Ability and Dyslexia*, ed. I. Padgett (London: Central Saint Martins College of Art and Design, 1999).
8. ddig.lboro.ac.uk/2004_conference/documents/sarah_parsons_notes.doc.
9. U. Wolff and I. Lundberg, The prevalence of dyslexia among art students. *Dyslexia* 8 (2002): 34–42.
10. T. G. West, *Thinking Like Einstein: Returning to Our Visual Roots with the Emerging Revolution in Computer Information Visualization* (Amherst, MA: Prometheus Books, 2004).
11. M. Wolf, *Proust and the Squid.*
12. www.iconeye.com/index.php?option=com_content&view=article&id=2714:dys lexia--icon-013--june-2004.
13. N. Geschwind, Why Orton was right.
14. M. Critchley and E. A. Critchley, *Dyslexia Defined* (Chichester, England: R. J. Acford, 1978).
15. The hippocampus plays an important role in many aspects of memory. In this case, it's our memory for where things are. As we'll see later in the book, the hippocampus also plays an important role in the kind of episodic memory abilities that Kristen so prominently displayed, and which play a major role in the reasoning skills of many individuals with dyslexia.
16. C. F. Doeller, C. Barry, and N. Burgess, Evidence for grid cells in a human memory network. *Nature* 463 (2010): 657–61.
17. MX's story appears in *Discover* online: http://discovermagazine.com/2010/mar/23-the-brain-look-deep-into-mind.s-eye.
18. J. Hadamard, *The Psychology of Invention in the Mathematical Field* (Mineola, NY: Dover Publications, 1954).

19. K. M. Jansons, A personal view of dyslexia and of thought without language. In *Thought without Language*, ed. L. Weiskrantz (New York: Oxford University Press, 2002).
20. Ibid. Einstein similarly described some of his mental imagery as being of a "muscular type." In J. Hadamard, *The Psychology of Invention in the Mathematical Field.*

Chapter 7

1. S. Dehaene, *Reading in the Brain* (see chap. 3, n. 3).
2. N. A. Badian, Does a visual-orthographic deficit contribute to reading disability? *Annals of Dyslexia* 55 (2005): 28–52.
3. www.iconeye.com/index.php?option=com_content&view=article&id=2714:dyslexia--icon-013--june-2004.
4. R. Fink, *Why Jane and John Couldn't Read—and How They Learned: A New Look at Striving Readers* (Newark: International Reading Association, 2006).
5. R. I. Nicolson and A. Fawcett, *Dyslexia, Learning and the Brain* (Cambridge, MA: MIT Press, 2010).
6. Persistent generation of mirror-image symbols—which results from the preservation of bilateral brain processing pathways—appears to be yet another example of how the slower acquisition of "mature" or "expert" processing in many individuals with dyslexia may lead to persistence of less mature and more bilateral (or bihemispheric) brain processing.
7. These brain regions include the planum temporale, supramarginal gyrus, and angular gyrus.
8. J. Hadamard, *The Psychology of Invention in the Mathematical Field.*
9. Two additional studies, one led by Vanderbilt language specialist Dr. Stephen Camarata and the other by Stanford economist Dr. Thomas Sowell, have also shown that severe late-talking is more common in children whose close family members work in "analytic" occupations. Many of these occupations, like engineering, scientific research, and airline piloting, are high M-strength professions. Both studies are discussed in T. Sowell, *The Einstein Syndrome: Bright Children Who Talk Late* (New York: Basic Books, 2002).
10. A. M. Bacon, S. J. Handley, and E. L. McDonald, Reasoning and dyslexia: A spatial strategy may impede reasoning with visually rich information. *British Journal of Psychology* 98 (2007): 79–82.

Chapter 8

1. See F. Epstein, *If I Get to Five: What Children Can Teach Us about Courage and Character* (New York: Holt Paperbacks, 2004).

Chapter 11

1. J. Everatt, B. Steffert, and I. Smythe, An eye for the unusual: Creative thinking in dyslexics. *Dyslexia* 5 (1999): 28–46.
2. J. Everatt, S. Weeks, and P. Brooks, Profiles of strengths and weaknesses in dyslexia and other learning difficulties. *Dyslexia* 14 (2007): 16–41.
3. We owe these interesting facts to Jeff Gray at the Gray-Area website (www.gray-area.org/Research/Ambig/).
4. These experts include Dr. Maryanne Wolf, Thomas G. West, and Dr. Albert Galaburda.
5. J. Lovelock, *The Revenge of Gaia* (New York: Basic Books, 2006).
6. T. R. Miles, G. Thierry, J. Roberts, and J. Schiffeldrin, Verbatim and gist recall of sentences by dyslexic and non-dyslexic adults. *Dyslexia* 12 (2006): 177–194.
7. Remarkably, we even see this kind of "upside surprise" in story comprehension in some of the children diagnosed with Specific Language Impairment, which shares many of the processing features of dyslexia but is associated with more severe difficulties comprehending language. These children typically have difficulties comprehending all but the shortest and most transparent sentences. When given a story with enough context and redundancy, these children often comprehend far better than expected. We've seen children score as much as three standard deviations higher on their oral story comprehension than on their vocabulary and single-sentence comprehension. This is likely due to their strengths in gist detection and top-down context-based processing.

Chapter 12

1. Also called *malapropisms.*
2. Thomas G. West has an interesting and insightful discussion of the phenomenon of paralexia in *In the Mind's Eye*, p. 43ff.
3. S. H. Carson, J. B. Peterson, and D. M. Higgins, Decreased latent inhibition is associated with increased creative achievement in high-functioning individuals. *Journal of Personality and Social Psychology* 85 (2003): 499–506.

Chapter 13

1. In reading, too, Douglas developed strategies. "When I could no longer get away with manipulating people, I built a bunch of tricks to try to get through reading. . . . It's mostly just modified skimming methodologies where you just sort of tag stuff as you go by, because there was no way I was going to be able to read things in any detail. But I could skim, and I would mark something like, 'This might be interesting later,' and then I would skim the thing over and over again, but I'd try not to think of it as reading, because if I thought of it as reading I'd get all worried about failing and how hard it was and I'd work myself into a frenzy, and that was not helpful. But by skimming over and over again and making progressively more organized marks, I could get the key elements out of an article, or a paper, or a chapter."
2. http://otherendofsunset.blogspot.com/.
3. This belief is beautifully exemplified by one of his projects at Google: "Imagine that people could ask questions of the world around them and get back answers that don't entirely match their perspective. How terrific would it be if it were possible for all of us to read what the Arabic newspapers were saying about our operations in the Middle East. How good would it be for the world if the democratization of information got to the place where consumers could see their own perspectives, the perspectives of those they trust and the perspectives of people who disagree with them all together and compare them." A. Lundberg, IT's Third Epoch . . . and Running IT at Google. *CIO* (2007). http://www.cio.com/article/144500/IT_s_Third_Epoch...and_Running_IT_at_Google.

Chapter 14

1. Interest in philosophy has been a common theme among the "dyslexic families" with whom we work. Earlier we mentioned that the most common college major among parents of our dyslexic children was engineering. The second-most common was philosophy. While only about 3 percent of college graduates in the United States majored in philosophy, over 12 percent of our parents with personal or family histories of dyslexia did. That's over four times the expected rate.
2. D. Seidman, *How: Why HOW We Do Anything Means Everything . . . in Business (and in Life)* (New York: Wiley, 2007).

Chapter 15

1. Both this and all other quotes attributed to Anne Rice have, unless otherwise stated, been taken from her autobiography, *Called Out of Darkness: A Spiritual Confession* (New York: Alfred A. Knopf, 2008).
2. We've posted a somewhat longer—but by no means exhaustive—list of these and other dyslexic writers on our Dyslexic Advantage website (http://dyslexicadvantage.com).

Chapter 16

1. Dr. Demis Hassabis has a curriculum vitae that sounds like it was dreamed up by Stan Lee as the backstory for a superhero. While he doesn't admit to slinging webs or turning green and muscley when he gets angry, he was a chess master at age twelve, won the world Pentamind championship at the Mind Sports Olympiad a record five times, and became a successful video game designer at age seventeen. He also earned a double first-class degree in computer science from Cambridge and started a successful video game production company with sixty-five employees—all by the time he'd reached his mid twenties. After successfully selling his company, he decided to combine his interests in imagination, creativity, and artificial intelligence by pursuing a Ph.D. in cognitive neurosciences at University College, London.
2. D. Hassabis, D. Kumaran, S. D. Van, and E. A. Maguire, Patients with hippocampal amnesia cannot imagine new experiences. *Proceedings of the National Academy of Sciences, USA* 104 (2007): 1726–31. See also D. Hassabis and E. A. Maguire, Deconstructing episodic memory with construction. *TRENDS in Cognitive Science* 11 (2007): 299–306.
3. From a test called the Boston Diagnostic Aphasia Examination. H. Goodglass and E. Kaplan, *Boston Diagnostic Aphasia Examination*, 2nd ed. (Philadelphia: Lea and Febiger, 1983).

Chapter 20

1. S. Andrews, Spatial thinking with a difference: An unorthodox treatise on the mind of the geologist. *AEG News* 45, no. 4 (2002), and 46, nos. 1–3 (2003).
2. In contrast to her poor verbal performance, Sarah excelled on the math portion of the SAT—despite being a C student in math class—prompting her math teacher to ask her, "Where have you been hiding this?" Sarah explained that the difference was entirely due to the SAT's multiple-choice format, which elimi-

nated her problems showing work and removed any penalty for her "original" way of doing math. Like many of the individuals we've mentioned in previous chapters, Sarah had difficulty memorizing and following the standard math formulas and procedures, so she created her own and did most of her work in her head. This led to conflict with her teachers. "My goal was, 'Let's get the right answer,' but theirs was, 'Let's do it the right way.'"

3. In a fascinating twist on this story, rather than work as a geologist, Sarah's aunt Lysbeth taught grade school and became a specialist in teaching what she termed "reluctant readers."

Chapter 21

1. J. Horner, *Dinosaurs under the Big Sky* (Missoula, MT: Mountain Press, 2001).
2. S. Andrews, Spatial thinking with a difference.
3. Ibid.

Chapter 22

1. S. Andrews, Spatial thinking with a difference.
2. Ibid.
3. M. Jung-Beeman et al., Neural activity when people solve verbal problems with insight. *Public Library of Science—Biology* 2 (2004): 500–510.

Chapter 23

1. Dr. Logan reported that the incidence of dyslexia is 20 percent among entrepreneurs in the United Kingdom, where the population incidence of dyslexia is estimated at 4 percent, and 35 percent of entrepreneurs in the United States, where the population incidence is around 10 to 15 percent.
2. Glenn Bailey also gave us a great example of the way that personal relationships can greatly affect worker satisfaction and performance. "When we ran our first water company, we had a great relationship with our team, and we had virtually no claims for work-related injury claims, despite the fact that we hand-delivered all these huge five-gallon bottles of purified water. When we sold the company, the people who bought us out were all about cash and bottom line, and they got rid of the Ping-Pong table and the barbecue and they got unionized, and their claims went through the roof. As a result, they became number one in worker injury claims in British Columbia."

Chapter 24

1. R. J. Bidinotto, Vince Flynn interview (2008). http://ayn-rand.info/ct-2066-vince_flynn.aspx.

Chapter 25

1. For example, many Orton-Gillingham–based programs, like the Wilson or Slingerland method, use fine-motor kinesthetic training, which emphasizes repeated practice in writing letters and words, or making tracing movements of the fingers. While these approaches are highly effective for individuals with dyslexia with good kinesthetic-spatial memories, individuals who are particularly weak in motor-kinesthetic imagery (and who often show significant problems with fine-motor finger coordination) often find these approaches both frustrating and ineffective, and they will learn better with programs that engage other areas of learning strength. Students with strong visual imagery but weak motor-kinesthetic imagery will often learn better with programs like Lindamood-Bell's Seeing Stars, which heavily engage visual imagery. Also, for students who struggle with finger coordination and position sense (i.e., the ability to tell what the fingers are doing without looking at them) but have good large-motor coordination and position sense, kinesthetic approaches that practice writing words and phonemes using large sweeping motions of the whole arm on a whiteboard rather than writing with the fingers using pencil and paper may also be effective. Other techniques that engage the visual, spatial, design, and color memory strengths in visual and spatial imagery that a particular student possesses will likely also be effective.
2. An example of this approach would be the Phonetic Zoo program, which couples phonics instruction with information about animals (see Appendix A).
3. B. L. Eide and F. Eide, *The Mislabeled Child* (see chap. 3, n. 1).
4. From a strictly neurological standpoint, there's a great deal of scientific evidence suggesting that many children who struggle to master fine-detail phonological skills may also have difficulty mastering fine-detail visual skills. This is even more likely for the large group of dyslexics who struggle with fine-detail movements of the fingers, which are required for tasks like handwriting and tying shoelaces.

Chapter 26

1. B. M. Vitale, *Unicorns Are Real: A Right-Brained Approach to Learning* (Austin, TX: Jalmar Press, 1982).

2. R. D. Davis and E. M. Braun, *The Gift of Dyslexia: Why Some of the Smartest People Can't Read . . . and How They Can Learn* (New York: Perigee, 2010). There are certain aspects of this book and of the Davis Method with which we are not fully in agreement, and many of his theoretical ideas seem far off the mark, but the practical material on building 3-D models of letters and words and on what Davis calls "trigger words" are often very helpful and are not well covered elsewhere.
3. D. Hanbury King, *Writing Skills* 1 and *Writing Skills* 2 (Cambridge, MA: Educators Publishing Service, 1990).
4. www.inspiration.com/blog/2011/01/discover-ways-to-showcase-dyslexic-talents/.
5. N. Margulies, *Mapping Inner Space: Learning and Teaching Mind Mapping* (Tucson, AZ: Zephyr, 1991).
6. This book, for example, was written entirely on a laptop, so handwriting played no role in its construction.

Chapter 27

1. M. H. Raskind, R. J. Goldberg, E. L. Higgins, and K. L. Herman, Patterns of change and predictors of success in individuals with learning disabilities: Results from a twenty-year longitudinal study. *Learning Disabilities Research and Practice* 14 (1999): 35–49.
2. M. E. P. Seligman, *The Optimistic Child: A Revolutionary Program That Safeguards Children against Depression and Builds Lifelong Resilience* (New York: Houghton Mifflin, 1995), and *Authentic Happiness: Using the New Positive Psychology to Realize Your Potential for Lasting Fulfillment* (New York: Free Press, 2002).

Chapter 28

1. B. L. Eide and F. Eide, *The Mislabeled Child* (see chap. 3, n. 1).
2. J. Mooney and D. Cole, *Learning Outside the Lines: Two Ivy League Students with Learning Disabilities and ADHD Give You the Tools* (New York: Fireside, 2000).
3. L. Pope, *Colleges That Change Lives: 40 Schools That Will Change the Way You Think about College* (New York: Penguin, 2006), and *Looking Beyond the Ivy League: Finding the College That's Right for You* (New York: Penguin, 2007).
4. Landmark also offers summer programs for high school students at various spots around the country. These programs run for two or three weeks, and they prepare students for intensive high school or college work by providing training in executive function, organization, learning strategies, and the use of assistive tech-

nologies. For students who are struggling with some of the skills that will be necessary to succeed in college, these courses can provide an excellent chance to acquire them.

Chapter 29

1. So far, this program has been very popular with mentors as well as mentees. One reason for this popularity is that the demands placed on the already busy mentors are quite reasonable. Mentors must commit to only twelve hours total, and they're given training and oversight from the organizers, so they always feel directed and supported.

INDEX